Understanding the CDM Regulations

D0756402

Also available from Taylor & Francis

Understanding the Building Regulations 3rd ed
S. Polley
Pb: ISBN 0–415–34917–6

OSH in Construction Project Management
H. Lingard et al.
Hb: ISBN 0–419–26210–5

Construction Project Management
P. Fewings
Hb: ISBN 0–415–35905–8
Pb: ISBN 0–415–35906–6

Practical Construction Management 2nd ed
R. Ranns et al.
Pb: ISBN 0–415–36257–1

Risk Management in Projects 2nd ed.
M. Loosemore et al.
Hb: ISBN 0–415–26055–8
Pb: ISBN 0–415–26056–6

Understanding the CDM Regulations

Owen Griffiths

Taylor & Francis
Taylor & Francis Group

LONDON AND NEW YORK

First published 2007
by Taylor & Francis
2 Park Square, Milton Park, Abingdon, Oxon OX14 4RN

Simultaneously published in the USA and Canada
by Taylor & Francis
270 Madison Ave, New York, NY 10016, USA

*Taylor & Francis is an imprint of the Taylor & Francis Group,
an informa business*

© Owen Griffiths 2007

Typeset in Sabon by RefineCatch Limited, Bungay, Suffolk
Printed and bound in Great Britain by
MPG Books Ltd, Bodmin, Cornwall

British Library Cataloguing in Publication Data
A catalogue record for this book is available from the British Library

Library of Congress Cataloging-in-Publication Data
A catalog record for this book has been requested

ISBN10: 0–419–22420–3 (pbk)
ISBN10: 0–203–96518–3 (ebk)

ISBN13: 978–0–419–22420–4 (pbk)
ISBN13: 978–0–203–96518–4 (ebk)

Contents

Figures

Tables

Acknowledgements

I would like to acknowledge the tireless work undertaken by Alun Griffiths without whose valued research and dedication this project would not have been possible. His passion for raising the standard of health, safety and welfare in our industry serves as a continual reminder as to why the principles of CDM must be embraced. Crown Copyright material is reproduced with the permission of the Controller of HMSO and the Queen's Printer for Scotland.

Introduction

This book has been written to provide students, CDM duty holders, health and safety consultants and construction professionals with practical advice on the Construction (Design and Management) Regulations 1994 (CDM) and their intended objectives. It appreciates that organisations and project duty holders need clear arrangements and tools for application, and the book also aims to provide guidance on designing appropriate and succinct management systems to facilitate compliance.

The theme of the book is that if CDM is to work it must be an integrated, indeed embedded, element of all the planning, design, management and construction functions. It is understood that professionals in the construction industry have many elements to manage at various stages throughout a project. CDM provides additional impetus for one more manageable element, namely the 'hazard'.

The benefits of an effective CDM strategy at a project level will also be covered, and the added value of such a policy will be illustrated with case studies throughout.

The book consists of seven chapters plus appendices. Chapter 1 consists of an overview of health and safety in the construction industry, legislation and the fundamental principles of the CDM Regulations themselves. As we develop this holistic picture of health and safety in the construction industry to date we shall hopefully prove the case for the necessity for appropriate project hazard management arrangements, and this will also allow us to move on to more detail in the subsequent chapters.

Chapter 2 covers the role of the CDM client and client's CDM agent. One could argue that without their commitment to risk management and without the client establishing clearly defined CDM project arrangements the effective management of health and safety is significantly weakened. We will also attempt to establish how the regulations cater for the diversity of client types and the wide disparity of client knowledge and how CDM caters for the respective procurement routes.

Chapter 3 covers the role of designers. They have the potential to initially impact on the health and safety of a project by virtue of their duty to inform the client of their CDM duties. Also, arguably, their greatest contribution is by their duty to design with adequate regard to health and safety. Early and ongoing intervention by a designer in designing to avoid or reduce the impact of environmental and social issues and hazards associated with certain materials and construction activities has significant benefits on the health and safety risk profile of a project. Recent surveys, however, commissioned by the Health and Safety Executive for example, have indicated that this potential is not being fully realised and that the design fraternity is failing to make a demonstrable contribution to the risk management process. The CDM Regulations also define a designer as more than the traditional architect or engineer, and this far-reaching definition will be explored.

Chapter 4 looks at the role and issues pertaining to the planning supervisor. The role was created by the UK lawmakers primarily to ensure design health and safety coordination and management. This requires skills of design, design management, construction processes *and* knowledge of occupational health and safety legislation. As well as addressing their statutory responsibilities we will debate their respective effectiveness since the introduction of the regulations and indeed look at the industry that has grown up around the planning supervisor.

Chapter 5 explores the duties of the principal contractor and the responsibility they have to manage all aspects of health and safety on site. The principal contractor is not always the main contractor on a construction project; we shall explore their statutory responsibilities and by means of case studies look at examples of good practice that exist to communicate and coordinate health and safety where it matters most – on site.

Chapter 6 looks at the role of the contractor and their working relationship with the principal contractor as a part of the site team. Contractors were facilitating health and safety legislation long before the introduction of CDM, and it is interesting to see how they have incorporated their CDM duties into their existing procedures.

Chapter 7 provides a complete CDM management system of all duty holders, including a useful project checklist. This will make the book more practically useful and set you on your way in terms of being able to actively apply the learning outcomes from the text.

As you read through the book the author implores you to develop your own thoughts on the CDM Regulations and their potential effectiveness to reduce fatalities, injuries and incidents of ill health. It is only through such individual opinions and debate that improvements are made in legislation as we wrestle with our social responsibilities knowing that, in the last 25 years, according to the Health and Safety Commission, over 2,800 people have died from injuries they received as a result of construction work and many more have been seriously injured or made ill.

I passionately believe in the importance of tackling the industry's health and safety problems. Pre-planned, well designed projects, where inherently safe processes have been chosen, which are carried out by companies known to be competent, with trained work forces, will be safe: they will also be good, predictable projects. If we are to succeed in creating a modern, world-class industry, the culture of the industry must change. It must value and respect its people, learn to work in integrated teams and deliver value for clients' money.

Sir John Egan, *Accelerating Change*, 2002

Chapter 1

Overview of the CDM Regulations

What are the CDM Regulations?

History of the CDM Regulations

As a member state of the European Union the UK is subject to legal directives, which it must interpret and implement. Directives impose a duty on each member state to:

- make regulations to conform with any directive; and
- enforce those regulations.

The Construction (Design and Management) Regulations 1994 are, therefore, the UK's response to EU Directive 92/57/EEC, 'The Management of Health and Safety Requirements at Temporary or Mobile Construction Sites'. Although, arguably, existing health and safety legislation was in place to cover many of the areas it was hoped CDM would address, such as the Health and Safety at Work etc. Act 1974 and the Management of Health and Safety at Work Regulations 1992 (now 1999), it was deemed necessary to provide additional impetus for construction safety. The regulations became effective on 31 March 1995 and marked a new era in construction health and safety management.

An initial Approved Code of Practice was published which attempted to offer practical advice and guidance on implementation of the regulations. The Health and Safety Executive also produced a series of Information Sheets on specific areas of CDM to assist the new legislation.

However, it is probably true to say that initial CDM responses were wide of their intended mark. Instead of the new legislation being seen as an attempt to further integrate health and safety management into the holistic project management plan, CDM was largely implemented as an overly bureaucratic 'bolt-on' that ran in parallel to the rest of the project.

Whether the CDM Regulations being largely misunderstood was down to a lack of clarity or guidance, a general lack of competence or a lack of desire by the respective industry bodies to embrace change remains a topic for debate. Significantly, the Health and Safety Executive considered a review of the Approved Code of Practice and the industry was asked to contribute to the process. Consequently, on 1 February 2002 a new Approved Code of Practice complete with additional guidance became effective. Many believe this was a major breakthrough in terms of providing information to assist with practical implementation. The emphasis on the production of relevant and succinct information also helped to promote the principles of appropriateness.

More recently, with many feeling that the regulations are still too far removed from their objective of adding value to the health and safety management process, a complete review of the regulations was initiated. The industry was consulted during 2005, with a re-branding of the regulations being proposed, including additional responsibilities for project clients, a more demonstrable contribution from the design fraternity and a general increase in levels of competence. These proposals will be further examined in Appendix 3.

It is no easy task to measure the effectiveness of the CDM Regulations since their introduction in the spring of 1995. If they are ultimately to be measured by the number of fatalities, injuries and incidents of ill health that have occurred in and as a result of construction work since their inception then there is a case to suggest that they have not proved as dynamic as was originally hoped. One could ask, however, what would have been the result on health and safety statistics if this additional legislation, namely the CDM Regulations, had not been introduced at all.

Health and safety in the construction industry

The construction industry is quite simply the largest industry in the UK and constitutes approximately 9 per cent of the United Kingdom's gross domestic product. It has been estimated that over 2 million people work in the construction industry.

It is also generally regarded as a relatively high-risk industry to work in. For example, Health and Safety Commission statistics remind us that:

- In the last 25 years over 2,800 people have died from injuries received as a result of construction work.
- In 2004/05, 33 per cent of all worker fatalities occurred in the construction industry.
- The number of major injuries to employees in construction was 4,098 in 2002/03.
- There were 1,874 deaths from mesothelioma in 2003 – similar to the number in the previous two years (there were 1,862 and 1,866 deaths in 2001 and 2002 respectively). Many of these deaths will be a result of past occupational exposures to asbestos.

There has been a major move by the industry itself to improve the appalling levels of fatalities, major injuries and incidents of ill health over the last few years. The first construction summit was held in February 2001 where the industry decided to set targets for reducing the rate of fatal and major injury to workers by 66 per cent by 2009/10. The summit was seen as an opportunity to raise the profile of construction health and safety and the bold targets set there showed an impressive intent. Other important initiatives include *Revitalising Health and Safety, Rethinking Construction* (the Egan Report) and *Achieving Excellence in Construction*.

The introduction of the CDM Regulations themselves must also be viewed as a major opportunity for the project team to focus further on the specific health and safety issues. The quotation below from John Egan is very much in line with the ideology that CDM is striving to achieve:

I passionately believe in the importance of tackling the industry's health and safety problems. Pre-planned, well designed projects, where inherently safe processes have been chosen, which are carried out by companies known to be competent, with trained work forces, will be safe: they will also be good, predictable projects. If we are to succeed in creating a modern, world-class industry, the culture of the industry must change. It must value and respect its people, learn to work in integrated teams and deliver value for clients' money.

Sir John Egan, Accelerating Change, 2002

Health and safety legislation in the UK

In order to appreciate the significance of the CDM Regulations in legal terms we will now, albeit briefly, look at the legal framework in the United Kingdom as well as our obligations as a member state of the European Union.

The Health and Safety at Work etc. Act 1974 (HSWA)

This is the primary source of health and safety legislation in the United Kingdom. It was introduced to unify existing legislation and raise awareness of occupational health and safety. It aimed to move the focus on to 'people at

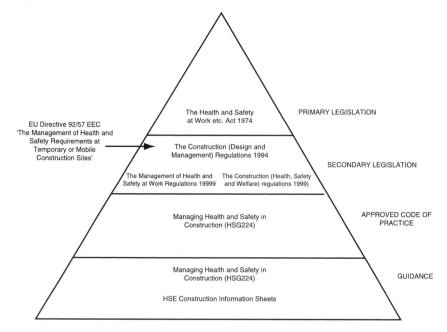

Figure 1.1 CDM legal framework.

work' in an attempt to reduce the general apathy that surrounded the subject of health and safety at the time.

The Act provided bold general duties for employers, the self-employed and even employees. For example:

> It shall be the duty of every employer to ensure, so far as is reasonably practicable, the health, safety and welfare at work of all his employees.
>
> HSWA section 2(1)

The Act also makes it necessary for employers to conduct their undertakings in such a way as to ensure, as far as is reasonably practicable, that persons not in their employment who may be affected thereby are not exposed to risk to their health and safety. The general public, for example, is a good example of a cohort whose health and safety an employer must consider.

Significantly, the Act also established both the Health and Safety Commission and the Health and Safety Executive. The former was established primarily to make arrangements to ensure the health and safety of people and work and those affected by that work. This is achieved through the proposal of new legislation and standards, instigating and conducting research and providing information and advice.

The latter has a role to advise the Commission but is more commonly involved in, and legally responsible for, the enforcement of health and safety legislation.

The Act also has the ability to make specific health and safety regulations, which serve to strengthen and support the principles of the primary legislation.

Regulations

The Health and Safety Commission will, from time to time, propose additional legislation to facilitate its work. This is generally achieved through a regulation that is brought into force under the Act. The Commission's proposal is made to, and introduced by, the appropriate government minister.

Regulations are seen as an appropriate means of setting new health and safety standards in a relatively quick timescale. A proposed regulation will automatically become law 21 days after submission to Parliament if no objections are made.

However, before the Commission proposes any new legislation it will consult with all interested parties, such as relevant industry bodies and trade unions, for example.

Regulations are also used as a tool to implement the requirements of EU directives, as indeed was the case with the CDM Regulations.

There are numerous other health and safety regulations that are potentially relevant to the construction industry over and above CDM. Effectively, it could be argued that CDM sits above many of these other regulations as a management platform for their successful implementation and consequent risk reduction. It would be of value for a construction professional to have a general grounding in the most relevant regulations.

Examples of health and safety regulations (that can be relevant to construction work) include:

- the Management of Health and Safety at Work Regulations 1999;
- the Construction (Health, Safety and Welfare) Regulations 1996;
- the Work at Height Regulations 2005;
- the Control of Substances Hazardous to Health Regulations 2002 (COSHH);
- the Control of Asbestos at Work Regulations 2002;
- the Lifting Operations and Lifting Equipment Regulations 1998 (LOLER);
- the Provision and Use of Work Equipment Regulations 1998 (PUWER);
- the Manual Handling Operations Regulations 1992;
- the Control of Noise at Work Regulations 2005;
- the Control of Vibration at Work Regulations 2005;
- the Health and Safety (First Aid) Regulations 1981;
- the Personal Protective Equipment Regulations 1992;
- the Reporting of Injuries, Diseases and Dangerous Occurrences Regulations 1995 (RIDDOR).

Approved Codes of Practice

If it is felt that further clarity and practical guidance are necessary to facilitate compliance with a regulation then it may be necessary to introduce an Approved Code of Practice (ACoP). It is important to understand the legal status of an ACoP. It is not a legal requirement to follow the information provided by any ACoP. However, failure to adopt the recommendations of the document in practice may be taken by a court in any criminal proceedings as evidence of failure to comply with the regulation to which it relates. Only if the defendant has demonstrated compliance with an alternative equally effective measure will this not count against them.

In terms of the CDM Regulations, an ACoP was introduced immediately to assist with initial compliance. Unfortunately, the publication, *Managing Construction for Health and Safety* (L54), was poorly received and did not really succeed in providing the practical guidance that the respective CDM duty holders were looking for. Eventually, the original ACoP was replaced and, in February 2002, a new publication was introduced titled *Managing Health and Safety in Construction* (HSG224). This new ACoP was well received by those in the industry who appreciated the significance of an Approved Code of Practice. HSG224 has done much to explain how those involved in construction health and safety can interpret the CDM Regulations.

Guidance notes

One of the main reasons for the relative success of the revised Approved Code of Practice was the fact that the publication also contained a wealth of guidance and good practice for CDM duty holders to consider as they strove for compliance with the regulations.

The Health and Safety Commission and the Health and Safety Executive issue guidance notes either alongside an ACoP or independently. They are aimed at providing further practical advice and suggestions and can be more informative.

Guidance notes carry no legal standing, but examples of case law do exist where the lack of adoption of guidance notes has been successfully used as evidence for the prosecution.

Over and above the CDM Regulations, the construction industry has been provided with a great deal of valuable guidance by the Commission and the Executive, much of which is available at little or no cost.

Enforcement of the CDM Regulations

The Health and Safety Executive (HSE) has a specific statutory responsibility of its own for the enforcement of health and safety law, including therefore the CDM Regulations. It has a number of inspectors whose primary objective is to encourage compliance with health and safety legislation. To this end the inspectors are awarded substantial powers to facilitate this aim and achieve improvements. These powers include:

1   Entering premises, at any reasonable time, if the inspector believes it is necessary to carry out any statutory provision, taking a police constable with them if deemed appropriate.
2   Providing advice or warnings.
3   Issuing an improvement notice. This will require any contravention to be remedied in a time that will be specified. Anyone who receives an improvement notice will be entered on to the notices database on the Health and Safety Executive website, which is for public consumption.
4   Issuing a prohibition notice. If an inspector believes there is, or is likely to be, a risk of serious personal injury they may issue a prohibition notice to cease an activity at any time unless or until the issue is resolved. Anyone who receives a prohibition notice will also be entered on to the notices database on the Health and Safety Executive website, which is for public consumption.
5   Prosecution in the criminal law courts.
6   Investigations of accidents or incidents. This includes being able to direct that any premises or area be left undisturbed, and taking photographs and measurements, and samples of any articles or substances including atmospheric samples.

7   Asking any person (the inspector has reasonable cause to believe to be able to provide information) to answer such questions as the inspector may consider appropriate.

Any breach of the CDM Regulations is deemed a criminal offence under statute law and can bring about judicial proceedings through the criminal courts. This is generally at a magistrates' court or crown court depending on the nature and severity of the alleged breach.

CDM duty holders and their responsibilities

The principles of the CDM Regulations are simple in that they define responsibility and respective duties throughout all phases that make up a construction project. They have been designed to promote a proactive and integrated hazard management process that is addressed and communicated via defined means of communication and information flow.

There are five CDM duty holders, namely:

- the client (or client's CDM agent);
- the designer;
- the planning supervisor;
- the principal contractor;
- the contractor.

The client

Without doubt the client holds the key to a project's success in terms of health and safety. They firstly have the opportunity to appoint the project team and, if all appointees are competent, cultured and resourced to address the issues likely to be involved, then health and safety management will truly become an integrated process of project management.

Secondly, the client has a major role to play in terms of providing relevant information to assist the team when making important design and planning decisions. Without information on the extent of existing services, ground contamination, asbestos and structural loadings for example, it becomes unfair to expect designers and contractors effectively to fulfil their respective roles.

There are obviously varying degrees of client knowledge about construction health and safety legislation and management. Less experienced clients will depend on the designer to inform them on their duties as far as the CDM Regulations are concerned. These clients will also require the planning supervisor to be proactive and helpful. If the client remains uncomfortable or not entirely competent with their CDM duties they may appoint a client's agent who will assume the statutory responsibilities for the regulations.

The more experienced, professional construction clients are looking to establish good health and safety management procedures and protocols from the earliest opportunity. So-called 'smart clients' appreciate the valuable contribution they can make by setting high standards through appointments and through the provision of relevant information at the time it is needed.

There also exists a further opportunity for the client to contribute to the health and safety management process. They are charged with ensuring that no work starts on site until a suitably developed construction phase health and safety plan has been produced. This effectively acts as a procedural gateway to ensure that appropriate health and safety management arrangements exist on site.

The main responsibilities of the client include the following:

1   Appoint a planning supervisor who is competent and adequately resourced to address the health and safety issues likely to be involved in the project.
2   Appoint a principal contractor who is competent and adequately resourced to address the health and safety issues likely to be involved in the project.
3   Ensure, when engaging any designers or contractors, that they are competent and adequately resourced to address the health and safety issues likely to be involved in the project.
4   Provide the project team, via the planning supervisor, with relevant health and safety information about the site and its surrounding environment.
5   Ensure that no work begins on site until a suitably developed construction phase health and safety plan has been prepared.
6   Keep the health and safety file available for inspection.
7   So far as health and safety is concerned, ensure appropriate arrangements are made to manage the project.

The designer

Anyone designing a structure or contributing towards the design of a structure has a duty to consider the health and safety implications of their choices in terms of avoiding risks to anyone carrying out construction work or cleaning work as well as anyone affected by this work.

Designers have immense power when it comes to avoiding or reducing the risks that arise day to day in construction and are legally bound to wield that power. Having said this, designers obviously have other influences to consider such as cost, aesthetics, building regulations, planning, environmental impact, party walls, etc.

Inevitably, designers assess the macro health and safety issues of structural form, loadings and materials, for example, but it has been suggested that they are not looking sufficiently at a systematic risk avoidance or reduction strategy for the people expected to construct and maintain their designs.

The term 'designer' in the CDM Regulations is far-reaching, and the scope of the designer goes beyond that of the traditional architect or engineer. The full extent of who could be perceived as having designer duties will be clarified in

Chapter 3. Suffice to say, when looking at who has designer duties under the regulations, that we need to consider the 'function' over the 'title'.

The main responsibilities of the designer include the following:

1   Take responsible steps to inform the client of their duties under the CDM Regulations.
2   Give adequate regard to the hierarchy of risk control when carrying out design work.
3   Ensure design includes adequate information about health and safety.
4   Cooperate with the planning supervisor and other designers associated with the project.
5   Ensure, when arranging for any designer(s) to prepare a design, that they are competent and adequately resourced for health and safety.

The planning supervisor

The EU directive on which CDM was based called for 'a co-ordinator of health and safety matters'. This duty holder was, therefore, created specifically as a result of the regulations themselves and indeed is often referred to as a 'creature of the regulations'. It was clearly intended that the planning supervisor be an integral part of the design team, focusing primarily on the health and safety implications of the design. The revised Approved Code of Practice clearly spelt out the main responsibility when stating that:

> Planning supervisors have to be satisfied that designs address the need to eliminate and control risks. . . .
>
> CDM Approved Code of Practice HSG224

The planning supervisor will also take any significant residual risk information provided by the designers and along with relevant information provided by the client prepare the health and safety plan.

Traditionally, the planning supervisor has suffered, unfairly in many cases, from a particularly bad press. A competent planning supervisor who is adequately resourced and appointed prior to any significant design having being undertaken can make a most valuable contribution to health and safety. However, a general lack of understanding of the role of the planning supervisor has led to ambiguity and consequently scepticism.

The main responsibilities of the planning supervisor include the following:

1   Ensure notification is submitted to the Health and Safety Executive.
2   If required, be in a position to give adequate advice to clients on the designer's competence and provision for health and safety.
3   Ensure, so far as reasonably practicable, that designs comply with regulation 13.

4 Take reasonable steps to ensure cooperation between designers.
5 Be able to give adequate advice to clients on the contractor's competence and provision for health and safety.
6 Be able to give adequate advice to contractors on the designer's competence and provision for health and safety.
7 Ensure that a pre-tender health and safety plan is prepared in good time.
8 Be able to advise the client on the suitability of the initial construction phase health and safety plan.
9 Ensure that a health and safety file is prepared for each structure.
10 Ensure that the health and safety file is delivered to the client.

The principal contractor

With the responsibility to manage health and safety throughout the construction phase, the principal contractor has numerous responsibilities under the CDM Regulations.

The principal contractor has emerged as a result of the CDM Regulations and it is important not to confuse the principal contractor with the main contractor or any other contractor. This duty holder is solely involved with the effective implementation of their responsibilities under the CDM Regulations.

Much of the work of the principal contractor relates to ensuring that the standards and rules set in the construction phase health and safety plan are adhered to on site. Their relationship with the contractors is pivotal to successful health and safety management during the construction period.

The main responsibilities of the principal contractor are as follows:

1 Ensure a health and safety plan is prepared for construction work, and is implemented, monitored and kept up to date.
2 Take responsible steps to ensure cooperation between the contractors.
3 Ensure compliance with the rules (if these are made).
4 Take reasonable steps to ensure that only authorised people are allowed on to the site.
5 Display a copy of the project notification prominently on site.
6 Provide the planning supervisor with information relevant to the health and safety file.
7 Give reasonable directions to contractors and monitor their work.
8 Make rules in the health and safety plan. If they are made, they should be in writing.
9 So far as is reasonably practicable, ensure contractors provide training and information to employees.
10 Ensure workers can discuss and offer advice and that there are arrangements for coordinating their views.
11 Ensure when arranging for any designer(s) to prepare a design that they are competent and adequately resourced for health and safety.

12 Ensure when arranging for any contractor(s) to carry out or manage construction work that they are competent and adequately resourced for health and safety.

The contractor

Contractors obviously have an important part to play in health and safety management as they are expected actually to carry out the construction work. Many are familiar with the principles of risk management as there is other legislation that requires employers, the self-employed and indeed employees to assess risks and adopt suitable control measures based on the principles of prevention.

As far as CDM is concerned they need to cooperate and work closely with the principal contractor in such a way as to consider and manage the health and safety implications of their work.

The CDM Regulations require that contractors do the following:

1 Cooperate with the principal contractor.
2 Pass to the principal contractor information which will affect health and safety, is relevant to the health and safety file or is relevant to RIDDOR.
3 Comply with the directions of the principal contractor and all the rules in the health and safety plan.
4 Provide relevant information and training to employees.
5 Ensure when arranging for any designer(s) to prepare a design that they are competent and adequately resourced for health and safety.

The above information is merely a brief overview of the five duty holders involved in the CDM Regulations and is aimed at providing a flavour of their respective responsibilities. Each role is explored in detail later in the book.

Application of the CDM Regulations

We will next look at the CDM Regulations in terms of when they are applied to construction work. It is also necessary to define two key terms whose application it is important to be able to assess, namely 'construction work' and 'structure'. We shall also assess when the Health and Safety Executive need to be informed of an intention to initiate a construction project. The term 'notification' is used to describe this requirement.

Interpretation

It is obviously important to appreciate when the CDM Regulations are actually applicable to a construction project or construction activity. Regulation 3 is concerned with the application of the regulations and states that:

Subject to the following paragraphs of this regulation, these Regulations shall apply to and in relation to construction work.

CDM regulation 3(1)

This said, it is clearly evident that a definition of the term 'construction work' is desirable, and this clarity is provided by regulation 2(1) in that:

'construction work' means the carrying out of any building, civil engineering or engineering construction work and includes any of the following –

(a) the construction, alteration, conversion, fitting out, commissioning, renovation, repair, upkeep, redecoration or other maintenance (including cleaning which involves the use of water or an abrasive at high pressure or the use of substances classified as corrosive or toxic for the purpose of regulation 7 of the Chemicals (Hazard Information and Packaging) Regulations 1993), de-commissioning, demolition or dismantling of a structure,

(b) the preparation for an intended structure, including site clearance, exploration, investigation (but not site survey) and excavation, and laying or installing the foundation of the structure,

(c) the assembly of prefabrication elements to form a structure or the disassembly of prefabricated elements which, immediately before such disassembly, formed a structure,

(d) the removal of a structure or part of a structure or of any product or waste resulting from demolition or dismantling of a structure or from disassembly of prefabricated elements which, immediately before such disassembly, formed a structure, and

(e) the installation, commissioning, maintenance, repair or removal of mechanical, electrical, gas, compressed air, hydraulic, telecommunications, computer or similar services which are normally fixed within or to a structure,

but does not include the exploration for or extraction of mineral resources or activities preparatory thereto carried out at a place where such exploration or extraction is carried out.

Extract from CDM regulation 2(1)

Given the above references to the term 'structure', again a clear definition is required. This too is provided by regulation 2(1), which offers that:

'structure' means –

(a) any building, steel or reinforced concrete structure (not being a building), railway line or siding, tramway line, clock, harbour, inland navigation, tunnel, shaft, bridge, viaduct, waterworks, reservoir, pipe or pipe-line (whatever, in either case, it contains or is intended to contain), cable, aqueduct, sewer, sewage works, gasholder, road, airfield, sea defence works, river works, drainage works, earthworks, lagoon, dam, wall, caisson, mast, tower, pylon, underground tank, earth retaining structure, or structure designed to preserve or alter any natural feature, and any other structure similar to the foregoing, or

(b)   any formwork, falsework, scaffold or other structure designed or used to provide support or means of access during construction work, or

(c)   any fixed plant in respect of work which is installation, commissioning, de-commissioning or dismantling and where any such work involves a risk of a person falling more than 2 metres.

Extract from CDM regulation 2(1)

When the CDM Regulations apply

Knowing the extent of the legal definition of 'construction work' and similarly understanding the scope of a 'structure' in terms of the law, an attempt can now be made to establish when the CDM Regulations are applicable and also when notification is required to be made to the enforcing authority.

So if the work is construction, as defined above, there are several factors that determine which of the CDM Regulations apply to the project. Significantly, where the full extent of the CDM Regulations does not apply then the designer's duties can apply in their own right.

Figure 1.2 explains when CDM is applicable.

In summary, however, the following criteria are to be considered:

* Is there demolition or dismantling? (See Appendix 4, 'Glossary of terms' for a definition.)
* Is the project duration anticipated to be greater than 30 days or involve more than 500 person days? (See 'Notifiable' in Appendix 4, 'Glossary of terms'.)
* Will there be five or more people on site at any one time?

If the answer to any of these questions is yes, then the full extent of the CDM Regulations applies, but the designer's duties apply on all projects irrespective of these factors.

When notification is required

The regulations require that the Health and Safety Executive is provided with some basic project details. It is the responsibility of the planning supervisor to ensure that this notification is undertaken. On a domestic scheme, where the CDM Regulations are not fully applied, it is left to the contractor to notify the enforcing authority.

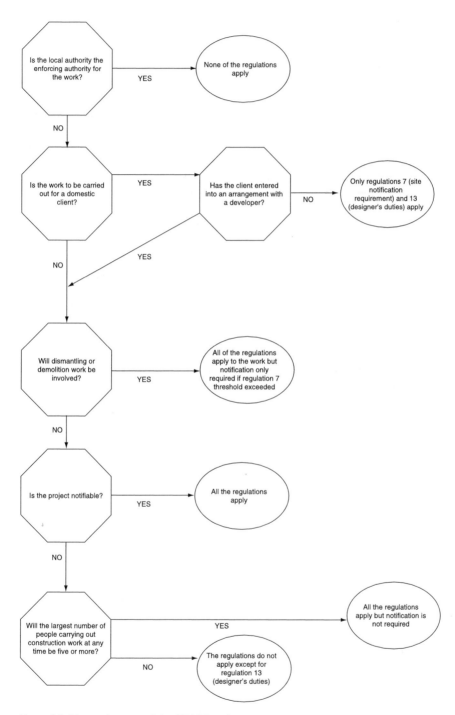

Figure 1.2 The application of the CDM Regulations.

Notification is based on the criteria illustrated in Figure 1.3.

Interestingly, many organisations have in-house polices of applying the CDM Regulations to *all* projects because they appreciate that designs need coordination, people should be competent and hazards must be managed effectively.

To help with compliance and policy implementation, a list of project types where CDM does, and does not, apply could form part of the client's CDM toolkit. This list complements the existing CDM application flow diagram (Figure 1.2), and together they can act as a valuable aid to demonstrate the consideration given to application of the regulations.

One important issue that we must consider when asking whether the CDM Regulations apply to a project is this. The CDM Regulations were introduced to implement a European directive, and they have been designed as management regulations for the appropriate management of hazards by defined project members. However, the Management of Health and Safety at Work Regulations 1999 (MHSW Regs) have all the legal elements to promote risk assessment, cooperation, competence and training, and provision of information. Therefore, if CDM is not applicable (with the exception of the designer's duties), then the MHSW Regs will be adequate to manage the hazards. Simply, if CDM does not apply then MHSW will.

Note: Some legal/CDM terminology can be misleading. Terms like 'shall', 'practicable' and 'reasonably practicable' are difficult to apply, even with a comprehensive and practical understanding. But to work within the regulations we need to understand what is expected of us and how far we need to go to execute a duty and comply. See Appendix 4, 'Glossary of terms' for guidance on these terms.

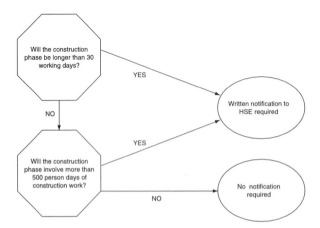

Figure 1.3 Project notification.

Project hazard management

If one of the objectives of CDM is hazard management then the primary function of all duty holders to make their health and safety contribution at the appropriate stage of the project is fundamental.

Hazard management concepts

The definition of a hazard is 'something with the potential to cause harm'. The management of problems (or hazards) should be considered by all professionals as a part of the management function and not something that just health and safety people do. Most decisions in construction have an impact on design but also have financial, business risk, regulatory and health and safety implications. One could argue that most of the above are related and to make decisions independently is not prudent.

To simplify, hazards could be thought of as 'problems' and the undertaking to address these as 'actions'. Problems can be associated with the following categories on a project:

- Existing environment, which includes:

   access to the site, e.g. busy roads, shared access, low bridges, restricted by size, etc.;
   access for materials to the workface, e.g. restricted by the adjacent building, overhead cables, etc.;
   ground conditions, e.g. contamination, wet, slopes, unstable buried structures, etc.;
   existing services, e.g. buried services, overhead services, etc.;
   materials, e.g. asbestos, lead paint, existing chemical storage, etc.;
   existing structures, e.g. unstable structures, existing temporary works, loose materials, etc.;
   local weather, e.g. flooding, high winds, exposed, etc.;
   existing activities, e.g. shops, operational use of the building, nearby school, etc.

- Design and construction, which include:

   manual handling, e.g. laying 140mm concrete blocks, 600 × 600 slabs, 86kg kerbstones, etc.;
   vibration and noise, e.g. using pneumatic machinery;
   falls from height;
   falling objects, e.g. during lifting operations;
   site traffic, e.g. large plant and machinery;
   hazardous materials;
   uncontrolled collapse, e.g. excavations, structural instability;
   confined spaces, e.g. tanks, voids, basements;

restricted access;
lifting operations, e.g. use of cranes;
existing services, e.g. buries services, overhead power lines;
deep excavations.

- Future maintenance, which includes:

  access for cleaning, e.g. working off ladders, tower scaffolding;
  component replacement, e.g. curtain walling, lights;
  access for maintenance, e.g. confined spaces, working at height.

If the primary objective of CDM is hazard management then an understanding of how valuable health and safety information is communicated through a construction project is required. Each CDM duty holder has a responsibility to provide and communicate information through a system of statutory protocols. We will now, albeit briefly, assess this holistic hazard management process. Figure 1.4 illustrates the process, which we shall next explore.

Client's hazard management contribution

A project, in terms of a hazard management concept, should start with the client identifying what information they need to provide to the planning supervisor and designers to enable them to consider the impact on the design and development of a tender package or construction brief. Providing the information to facilitate future consideration by a designer or contractor is a significant contribution towards managing the hazard.

The information a client should provide is based on 'reasonable enquiries', and examples of what they should provide is detailed in Chapter 2.

On many projects this work of information gathering is undertaken by the lead consultant, who is generally an architect or engineer. It is advisable to have a formal arrangement in place between the client and the person carrying out the work to define who is responsible for what.

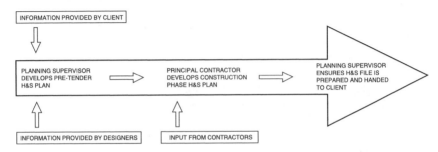

Figure 1.4 The holistic CDM flow of hazard management information.

Designer's hazard management contributions

Potentially the designer is in the best position to manage significant hazards based on their early intervention on the project and by the relationship between all the aspects that define the design and hazards. Early decisions about the location of a structure, the materials and processes and the fundamental design of the structure affect buildability and health and safety throughout the life cycle of the structure. The management of these problems must be integrated within the design process and promote the identification of hazards at all stages of design and initiate a design response that is weighted according to the level of risk. This is termed the control hierarchy. It promotes a design to avoid or eliminate significant hazards; where this is not possible the design needs to reduce the level of risk to where it is not significant; and if this is not possible suitable control measures are required or the hazard needs to be transferred to the contractor. The most common process for achieving this is risk assessment, which is discussed in detail in Chapter 3.

Designer's duties are unique in that they apply to all projects irrespective of the size, risk profile, number of persons working on the project or even if the project is for a domestic client. On a project where the full sets of CDM Regulations apply, the designer should communicate the outputs of their risk assessment to the planning supervisor.

Planning supervisor's hazard management contributions

The first contribution to be made by the planning supervisor is the collation of relevant information from the client on the site's environmental/inherent hazards. This can involve the planning supervisor or a competent person undertaking a site visit and recommending investigations to the client.

The second and probably most important contribution is the planning supervisor's duty to ensure designers design with adequate regard to health and safety and provide adequate information with their designs.

The significant residual risks from the site and the design are then collated into the pre-tender health and safety plan. This document alerts the principal contractor of the issues and allows the principal contractor to plan and resource for health and safety.

Finally the planning supervisor's duty to ensure that the health and safety file is suitably developed and provided to the building/structure manager is also significant. The health and safety 'file' holds valuable health and safety information for all future projects, maintenance activities and operations throughout the life of the building or structure.

Principal contractor's hazard management contributions

The principal contractor's appointment and position on the project as the manager of all aspects of health and safety during the construction phase places them in a pivotal position to contribute to health and safety management. The principal contractor should manage all aspects of safety through a project-specific set of arrangements called the construction phase health and safety 'plan'.

The standards and tools for the management of health and safety on site delivered through the construction phase health and safety plan include risk assessments, method statements, fire, emergency, site traffic management and lifting plans, training, including induction, registers for inspections and permits to work.

The principal contractor and contractor's hazard management role is at the sharp end of construction and has direct links with fatalities and ill health, but as professionals their early contributions as site safety managers and in organisations with knowledge of constructability are invaluable.

Risk assessment

Risk assessment is a discipline that all construction professionals must either legally or practically be confident in practising. From the earliest decisions on a project, the principles of risk assessment should be applied.

It is important to clarify that risk assessment is an integrated mindset that involves identification of problems and the development of solutions. The methodology, management of outputs and review should all be defined in either a health and safety policy or a safety plan before starting the process. The risk assessment pro forma should be considered as a report on one's findings that can be communicated for collaborative working, proof of compliance or action/implementation of safety or design standards depending on the objective.

Regulation 3 of the Management of Health and Safety at Work Regulations 1999 requires employers to undertake risk assessments of their activities. However, the CDM Regulations do not specify the need for any risk assessment, but this is the most common process by which designers demonstrate they are designing with adequate regard to health and safety.

The fundamental elements of risk assessment involve the process or mechanism, which should be different for different objectives, and the competence of the person undertaking the process in knowing what outputs or standards to define.

The management of actions through the risk assessment document can be expanded to the use of a common risk register that provides a facility that is managed by controllers and contributed to by a defined user group. The controller could be the planning supervisor for a design risk register or a project manager for an all-inclusive risk register, and the contributor could be the design team or the whole project team. Modern information technology, particularly the Internet, provides opportunities for project teams to run live risk registers that not only record the problems and solutions but also project-manage the actions and provide all the associated information to whoever requires it.

Hazard checklists

To start a risk assessment process, the hazard checklist can be beneficial. Examples of where such a checklist could be used on a project include:

- a developer making initial enquiries about a site;
- a client assessing the potential interface/impact of a project on existing operations, e.g. a highway, school/university, hospital, office, bridge, etc., and deciding what information to provide to potential designers and contractors;
- a designer undertaking a design review to check for significant hazards that could be designed out or the risk reduced through design and what information to provide to the contractor;
- a designer undertaking a post-construction activity assessment for large-component replacement and access for future maintenance including cleaning;
- a contractor undertaking an initial hazard identification process on a new project and allocating hazard ownership and actions to the contracting team.

The problems can be identified using the checklist as an aide-memoire or pro forma and recorded for action. Some of the hazards that could be considered on a checklist include:

- asbestos;
- falls from height;
- falling objects;
- fire risk;
- manual handling;
- noise and vibration;
- site traffic;
- highway traffic;
- railways;
- buried services;
- overhead services;
- confined spaces;
- existing activities on site;
- interface with the public;

- hazardous dust;
- large difficult-to-manage components;
- contaminated ground conditions;
- buried structures;
- condition of the existing structure;
- uncontrolled collapse;
- access restrictions (to site and workface);
- nearby/on-site water courses;
- lifting operations.

Design risk assessment pro forma

It could be argued that the design risk assessment pro forma is one of the most misused documents in the industry. The objective of the pro forma is to promote a process of design risk assessment, and not construction risk assessment, and to record and communicate with the planning supervisor and other designers as necessary.

When deciding what to record and what not to record on a design risk assessment the designer should consider the following:

- Is the hazard significant?
- Do I need to record a design action and thought process when the evidence is on the drawing?
- Would I expect a competent contractor to know what is recorded?
- Am I assuming the contractor will use a certain method of work?
- Would the contractor be aware of the asbestos report?
- Are the other designers aware and will they consider the problems with the site?

The design risk assessment pro forma has the potential to facilitate effective hazard management during the design phase, which on many projects goes on throughout the construction process. However, if not used effectively, such pro formas can facilitate an overly bureaucratic approach which can obscure the significant health and safety issues.

The health and safety plan

The health and safety plan is the medium used to articulate and communicate significant health and safety information throughout a project until the end of the construction phase. Prior to work beginning, during the tender stage for example, the contractors can view the plan and fully assess the significant health and safety issues and consequently ensure they can be sufficiently resourced and managed.

(1)   The planning supervisor appointed for any project shall ensure that a health and safety plan in respect of the project has been prepared no later than the time specified in paragraph (2). . . .

(2)   The time when the health and safety plan is required by paragraph (1) to be prepared is such time as will enable the health and safety plan to be provided to any contractor before arrangements are made for the contractor to carry out or manage construction work.

Extract from CDM regulation 15

Note that the regulation only requires the planning supervisor to *ensure* that a health and safety plan is prepared. There is therefore no statutory duty for the planning supervisor actually to prepare the document. Generally, however, it is indeed the planning supervisor who collates and prepares the health and safety plan pre-construction.

The principal contractor must develop the health and safety plan throughout the construction phase in that they need to consider and demonstrate how significant residual risk will be resourced and managed.

The Approved Code of Practice clearly divides the health and safety plan into two distinct entities, namely the pre-tender plan and the construction phase plan.

It is noteworthy, when emphasising the significance of the health and safety plan, that all project members make contributions to the document. Clients provide information on the project's inherent hazards; designers provide details of risks associated with their design; planning supervisors may offer additional information; and the principal contractor evolves and develops the plan complete with contributions from the contractors.

Pre-tender health and safety plan

The objective of the pre-tender (or pre-start) health and safety plan is to provide a project-specific and concise compendium of health and safety-related information upon which the principal contractor can base the strategy for construction health and safety management. The document is fundamental to the CDM process, as it contains information based on the following criteria:

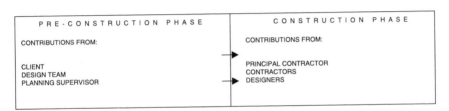

Figure 1.5 The health and safety plan.

- a description of the project, contact details and existing records and plans;
- the client's considerations and management requirements for the project, including site rules, existing emergency procedures, activities adjacent to and on site, and security arrangements during construction;
- environmental restrictions and existing on-site risks, which should include safety and health risks;
- significant design and construction hazards, including design assumptions and control measures, significant design risks including hazardous materials, and arrangements for coordinating ongoing design work and design changes;
- the format, content and arrangements for development of the health and safety file.

This plan also has the facility to set minimum health and safety standards, which, following the letting of the contract or start of work, would be difficult to impose thereafter on any contractor.

As the information and standards are important to the client's operation and investment, the assessment of the contractor response to the plan is equally important. The plan should take advantage of CDM regulation 10 as a legal procedural gateway in informing the contractor that the client cannot allow work to start until a suitably developed construction phase plan has been developed. It is the client's duty to assess this, but the planning supervisor must offer advice if asked.

Construction phase health and safety plan

The construction phase health and safety plan is the master control document for site safety. Its development for work to start must be based on the first stages of the construction programme but also should have the infrastructure to demonstrate that health and safety will be managed throughout the project. To this end, the development of the plan is critical as a hazard management tool. Regulation and assessment of the plan are legally completed by the client, often with advice from the planning supervisor.

A typical construction phase health and safety plan, if it is to be deemed suitably developed, should contain the following:

- a description of the project, contact details and existing records and plans;
- arrangements for the communication and management of the work, including liaison between parties on site, consultation with the workforce, the exchange of design information, risk assessment and method statement production and approval, site rules and site emergency procedures;
- arrangements for controlling significant site health and safety risks including temporary services, falls from height, lifting operations, traffic management, hazardous substances, removal of asbestos, contaminated land, reducing noise and vibration, manual handling, etc.;

- the health and safety file layout and format, arrangements for collection of file information and storage arrangements.

The construction phase plan can be a stand-alone document and on more complex projects it can be an integrated element of the organisation's site safety management system.

The health and safety file

From a risk profile perspective, given that the majority of construction and maintenance operations will be undertaken post-completion and indeed throughout the lifetime of the structure, the health and safety file is considered to be an invaluable source of information. One of the initial concepts of the health and safety file is that any additional work including demolition on the building or structure will be supported by the file's information.

The health and safety file could be useful post-project completion for:

- designers and contractors on future refurbishment projects;
- developing facilities maintenance or management contracts;
- electrical maintenance and upgrades to the system;
- mechanical maintenance and upgrades to the system;
- cleaning contractors planning work;
- the client or planning supervisor in developing a tender;
- an engineer tracing a fault on any system.

The file can be a stand-alone document or reference source but is often an integrated part of the operational and maintenance manual for the structure or building.

The biggest problem in developing a file is data management. The first stage is getting any data at all. The data then needs to be valuable. Irrelevant information should be avoided. The file must also be kept up to date with any amendments. Version control is important where multiple copies exist.

The format of the health and safety file should be based on the following information:

- A brief description of the work carried out.
- Residual risks and information to support their future management, e.g.

  asbestos surveys and clearance certificates;
  existing buried services layout;
  contaminated land report.

- Key structural principles within the design, e.g.

  sub-structural design, e.g. piles, foundations, etc.;
  structural frame/superstructure elements, e.g. bracing, floor loadings, roof

loadings, energy-stored elements like pre-tensioned and post-tensioned beams, etc.

- Hazardous materials used in the construction, e.g.

  lead paint on reclaimed windows;
  low-level contaminated land during some landscaping;
  finishes used in which there is a risk when repairing or cleaning.

- Future movement and dismantling of plant and equipment, e.g.

  load weights and lifting arrangements;
  decommissioning arrangements;
  method statements for removal/installation of difficult-to-manage operations;
  associated services information for removal/installation of difficult-to-manage operations.

- Equipment provided and arrangements for cleaning, e.g.

  access equipment training and safe operation;
  maintenance arrangements for equipment.

- Detail on the nature and location of significant building, system or structure services, e.g.

  fire-fighting mains;
  electrical supply;
  water supply;
  gas supply;
  telecommunications supply;
  sewage connections.

- As built and installed information on the structure and its equipment, e.g.

  substructure;
  main superstructure;
  roofing system;
  external finish;
  windows and atria;
  internal finishes;
  mechanical and electrical (M&E) details;
  fire doors and compartmentation;
  interface with existing structures.

The development of the file is generally completed at the end of the project, with the client and planning supervisor defining the arrangements for its completion. However, many clients and contractors alike are seeing advantages in the ongoing development of the file from the start of the design stage, owing to the valuable nature of the information throughout the construction phase.

Chapter 2

CDM clients and the client's agent

Introduction

It has long been argued that the client holds the key to the successful health and safety management of any construction project. In terms of the CDM Regulations this cannot be denied. Through the appointment of a competent, adequately resourced team and by allocating sufficient time for the project to be properly planned, designed and constructed, the client can truly make the maximum contribution to the health, safety and welfare on site.

The cultured client will also be aware of what information the team will require to be able to facilitate a proactively planned scheme. Designers and contractor need to be made aware of relevant health and safety information so they can make the right choices based on the level of risk.

There is also a critical procedural gateway whereby the client must ensure that a suitably developed construction phase health and safety plan has been compiled by the principal contractor. Only then can the client allow construction work to begin.

All too often though, the client has little or no understanding of the significant opportunity they hold to contribute to the health and safety management process. Clients can make the mistake of assuming that once they have appointed the team they can stand back from any health and safety responsibility yet still drive the project from all other angles.

The uneducated client who has not fully understood their role in a project's health and safety management can suffer potentially serious business risk issues by:

- enforcing unrealistic project timescales;
- appointing incompetent and under-resourced designers, planning supervisors and contractors;
- not providing relevant health and safety information when it is needed;
- allowing work to proceed without a suitably developed health and safety plan;
- not making the health and safety file available post-construction.

As many clients are not especially familiar with construction health and safety legislation, the CDM Regulations require designers to ensure that they are aware of their duties. The planning supervisor should also be able to advise the client on the health and safety aspects of appointments and the suitability of the construction phase health and safety plan and even help with the general project health and safety management arrangements.

If the client is still uncomfortable with their CDM responsibilities they may empower a client's agent to act on their behalf. This competent agent then assumes the client's CDM responsibilities. It has been suggested, however, that if a client simply passes on their legal health and safety duty by appointing an agent but continues to drive the project in all other ways, such as timescales and finance for example, the potential for CDM to work could be compromised or at least diluted. It may well be the case that the removal of the client's CDM agent will be an option when the CDM Regulations are next reviewed.

Who is the client?

Before we take a detailed look at the respective duties of the client it is worth defining who the client is in the terms of the regulations. Although it may appear obvious in many cases it is important to establish who, on a particular project, fulfils the CDM functions of a client.

Interpretation

The interpretation offered by the Regulations themselves states that:

> Client means any person for whom a project is carried out, whether it is carried out by another person or is carried out in-house.

Clients can come from a variety of organisations or individuals, including for example:

- housing associations;
- local authorities;
- companies;
- developers;
- NHS trusts;
- financial institutions;
- charities;
- government agencies.

Guidance from the CDM Approved Code of Practice (ACoP) suggests that other factors to consider, when it may not be immediately clear who the client is, include:

1. Who is at the head of the procurement chain
2. Who arranges for the design work, and
3. Who engages the contractors

ACoP HSG224 para 57

Where ambiguity exists, arrangements for carrying out the client's respective CDM duties will need to be established very early on. Lack of clarity exists, for example, where a large client organisation does not have effective health and safety policy arrangements or quality assurance procedures.

Question: A school is to build a large extension to facilitate a new sports complex. Who is the client in terms of complying with the CDM duties? Is it perhaps:

- the head teacher of the school?
- the board of governors?
- the chief executive of the local authority?
- the education department?
- building services?

The answer is that someone competent needs to fulfil the various client duties, and the arrangements for this undertaking should be clearly defined in the organisation's health and safety policy arrangements or perhaps integrated into project QA procedures.

Note that the CDM Regulations *do not* apply to a domestic client.

Developers

It is evident that all developers are clients in terms of the CDM Regulations.

However, the regulations also make specific mention of developers remaining the CDM client once they have sold to or entered into an agreement with a domestic buyer prior to the completion of a project but where they continue to arrange for construction work to be carried out. Regulation 5 offers:

(1) This regulation applies where the project is carried out for a domestic client and the client enters into an arrangement with a person (in this regulation called 'the developer') who carries on a trade, business or other undertaking (whether for profit or not) in connection with which –

(a) land or an interest in land is granted or transferred to the client; and

(b) the developer undertakes that construction work will be carried out on the land; and

(c) following the construction work, the land will include premises which, as intended by the client, will be occupied as a residence.

(2) Where this regulation applies, with effect from the time the client enters into the arrangement referred to in paragraph (1), the requirements of regulations 6 and 8 to 12 shall apply to the developer as if he were the client.

CDM regulation 5

Example: A property developer intends to build 30 new dwellings on a greenfield site. He is fortunate enough to sell 25 of the properties before he begins on site. As mentioned earlier, the CDM Regulations do not apply to domestic clients. However, in this instance the developer will remain a CDM client for the entire project as it is he who is undertaking or instigating the construction work.

Client's CDM duties explained

Making the appointment of a competent and adequately resourced team

This duty provides the client with an opportunity to make an immediate contribution to health and safety for the entire project from the very outset. They are the only duty holder who can appoint the planning supervisor and the principal contractor. In many cases the client is also responsible for the appointment of the lead designer.

Clearly, if the health and safety management of a project is to be in any way effective then the appointed team must be both competent and adequately resourced to address the health and safety issues likely to be involved.

Note that when we discuss competence and resources we are solely interested in these in terms of the ability to comply with the CDM obligations of the respective duty holder.

Competence of planning supervisors, designers and contractors

(1) No client shall appoint any person as planning supervisor in respect of a project unless the client is reasonably satisfied that the person he intends to appoint has the competence to perform the functions of planning supervisor under these Regulations in respect of that project.

(2) No person shall arrange for a designer to prepare a design unless he is reasonably satisfied that the designer has the competence to carry out that design.

(3) No person shall arrange for a contractor to carry out or manage construction work unless he is reasonably satisfied that the contractor has the competence to carry out or, as the case may be, manage, that construction work.

Extract from CDM regulation 8

Provision for health and safety

(1) No client shall appoint any person as planning supervisor in respect of a project unless the client is reasonably satisfied that the person he intends to appoint has allocated or, as appropriate, will allocate resources to enable him to perform the functions of planning supervisor under these Regulations in respect of that project.

(2)   No person shall arrange for a designer to prepare a design unless he is reasonably satisfied that the designer has allocated or, as appropriate, will allocate adequate resources to enable the designer to comply with regulation 13.

(3)   No person shall arrange for a contractor to carry out or manage construction work unless he is reasonably satisfied that the contractor has allocated or, as appropriate, will allocate adequate resources to enable the contractor to comply with the requirements and prohibitions imposed on him by or under the relevant statutory provisions.

CDM regulation 9

Traditional methods used by clients to assess the competence and resources of their proposed project team include the use of pre-qualification questionnaires and interviews. Generally, if an organisation has already established the competence of a particular principal contractor, for example from a previous project, then it may be possible to focus just on the adequacy of resources. There are dangers, however, associated with this philosophy as obviously no two projects are the same, personnel change and technologies move on.

The client must also have the ability to assess who is competent and who is adequately resourced. The development of a checklist to facilitate such assessments is sometimes undertaken.

It is, however, widely accepted that CDM competence is not what it should be, and improvements in the team's ability to understand and implement health and safety management philosophies throughout a project remains a major industry challenge.

Appointing the planning supervisor

It has long been argued that late appointments of relatively incompetent planning supervisors have been a contributory factor of inadequate health and safety planning from concept through to detailed design. Clients have not been made sufficiently aware of the valuable contribution that a competent planning supervisor can make. Their early appointment is a key CDM duty.

(1)   Subject to paragraph (6)(b), every client shall appoint –

(a)   a planning supervisor . . .

in respect of each project.

(3)   The planning supervisor shall be appointed as soon as is practicable after the client has such information about the project and the construction work involved in it as will enable him to comply with the requirements imposed on him. . . .

Extracts from CDM regulation 6

If the planning supervisor is to be in a position to advise on the competence and resources of the designers this naturally implies that they must be appointed at a very early stage. In fact, this is ideally as soon as the client is in a position to assess the competence and resources required of a planning supervisor.

The role of planning supervisor can be undertaken by an individual, an organisation or another project member or indeed can be an in-house appointment. What is required is for the client to ensure that the appointee is competent and adequately resourced to provide the service.

It is also worth noting that there is no such creature as a 'qualified' planning supervisor; therefore careful consideration is necessary to ensure the requirements of the appointment are satisfied.

The planning supervisor will require sufficient knowledge of the design processes involved and the construction activities likely to be adopted and must also have a sound understanding of occupational health and safety legislation. If these criteria are clearly apparent then competence should be relatively straightforward for the client to establish.

The most significant resource for a client to assess from a planning supervisor is arguably the time they are able to commit to the project. It is important for the client to know, for continued health and safety management continuity, that the planning supervisor is adequately resourced.

One anomaly with planning supervisors thus far has been the sometimes large inconsistency in the fees they request for their services. Whenever a client is assessing a fee proposal from a planning supervisor who they are considering appointing it is imperative to know exactly how far the service extends. In particular, clients with little CDM knowledge may find it difficult to understand why large variations exist between quotations for planning supervision on a project. With many of the planning supervisor's duties the planning supervisor must simply 'ensure' that a task is undertaken. Herein lies the relative inequality of fee proposals.

The planning supervisor must:

> ensure that a health and safety file is prepared in respect of each structure comprised in the project . . .
>
> Extract from CDM regulation 14(d)

Example: On a large construction project one planning supervisor quoted £500 to carry out the above statutory duty. A second planning supervisor quoted £5,000 for the same task as part of their fee proposal. You may well ask who the client is favouring thus far. However, the planning supervisor quoting £500 will ensure only that the health and safety file is prepared in that he will have the principal contractor actually put the file together. The other will request appropriate information and physically undertake all the necessary work to produce the file.

A competent planning supervisor is undoubtedly a valuable asset to any construction project, and the client must be educated to appreciate the added value the planning supervisor can bring to the health and safety management process.

Appointing designers

The client has no legal duty under the CDM Regulations to appoint a designer or lead designer. However, in many cases the designer is the first appointee made by the client. Consequently the regulations below will become effective in establishing their ability to comply with their duties:

> No person shall arrange for a designer to prepare a design unless he is reasonably satisfied that the designer has the competence to prepare the design.
>
> CDM regulation 8(2)

> No person shall arrange for a designer to prepare a design unless he is reasonably satisfied that the designer has allocated or, as appropriate, will allocate adequate resources to enable the designer to comply with regulation 13.
>
> CDM regulation 9(2)

Designers too, it has been suggested, are rarely challenged by clients in terms of proving that they are indeed competent to address the health and safety issues likely to be involved in their designs. More recently there has been a trend for clients to ask design practices to complete lengthy CDM pre-qualification questionnaires, but there is no evidence to suggest that this results in safer designs. A more practical suggestion for the assessment of designer CDM competence has been offered by the Construction Industry Council which requests simple, practical evidence of competence.

Appointing the principal contractor

Many clients understand the relevance of assessing the health and safety competence of contractors and it must be said that price is not always the overriding criterion to the successful award of a contract. Professional construction clients who appreciate the concept of a holistic business risk approach to project management will carefully consider the contractor's ability to manage health and safety on site. When making an appointment the client must consider the following:

> The client shall not appoint as principal contractor any person who is not a contractor.
>
> CDM regulation 6(2)

The principal contractor shall be appointed as soon as is practicable after the client has such information about the project and the construction work involved in it as will enable the client to comply with the regulations imposed on him by regulations 8(3) and 9(3) when making an arrangement with a contractor to manage construction work where such an arrangement consists of the appointment of the principal contractor.

CDM regulation 6(4)

Generally, good practice would demand that the competence of a principal contractor is assessed prior to an invitation to tender. This is sometimes undertaken with pre-qualification questionnaires. The establishment of adequate resources is often then undertaken at the tender stage.

There will only be one principal contractor on a project at any one time, but it is possible for a change in the appointment, for example if there is a significant change in the project's nature or perhaps if the principal contractor is clearly not able to carry out their CDM duty because of a decrease in competence or the provision of insufficient resources.

Providing relevant health and safety information to the team

It is the responsibility of the client to provide the team, usually via the planning supervisor, with relevant information relating to the proposed work. The client shall ensure the information is made available:

as soon as is reasonably practicable but in any event before the commencement of the work to which the information relates . . .

Extract from CDM regulation 11(1)

This may sound like an obvious statement, but clients have a duty proactively to provide details on the condition of any proposed site. The term 'reasonable enquiries' is used to define the extent of detail required and this is supported by the following list of examples in the CDM Approved Code of Practice (ACoP):

(a) the presence, location and condition of hazardous materials, such as asbestos or waste chemicals;
(b) activities on or near the site, which will continue during construction work, e.g. retail shops, deliveries and traffic movements, railway lines or busy roads, public access to a retail store;
(c) requirements relating to the health and safety of the client's employees or customers, e.g. permit-to-work systems in a petrochemical plant, fire precautions in a paper mill, one-way systems on site, means of escape, 'no-go' areas, smoking and parking restrictions;
(d) access and space problems, such as narrow streets, lack of parking, turning or storage space;

(e) information about means of access to parts of the structure, e.g. fragile storage space; materials and anchorage points for fall arrest systems;

(f) available information about site services and their location, in particular, about those that are concealed, such as underground services;

(g) ground conditions and underground structures or water courses, such as culverts, where this might affect the safe use of plant, e.g. cranes, or the safety of groundworks, e.g. the construction of trenches;

(h) buildings, other structures or trees which might be unstable or at risk of uncontrolled collapse;

(i) previous structural modifications, including weakening or strengthening of the structure;

(j) fire damage, ground shrinkage, movement or poor maintenance which may have adversely affected the structure;

(k) any difficulties relating to plant and equipment in the premises, such as overhead service gantries whose height restricts access;

(l) health and safety information contained in earlier design, construction or 'as built' drawings, such as details of pre-stressed or post-tensioned structures.

This provision of information is the first hazard management process on a project. It begins the flow of information and hazard management action by the initial members of the team.

It is questionable as to whether the design team and the planning supervisor can effectively carry out their duties if relevant health and safety information is not forthcoming from the client. For example, if a structural engineer is not made aware of ground contamination how can they confidently design with adequate regard to health and safety in terms of considering appropriate piling technologies?

Infinite other examples exist of designers, planning supervisors and contractors needing to be provided with relevant information to consider if they are to effectively carry out their duties.

Many clients already have some information, for example in existing health and safety files, and relevant surveys may have been undertaken. It is prudent also to discuss with the planning supervisor additional information that it would be useful to provide.

A hazard checklist could be employed as an aide-memoire to focus the client's attention on the more relevant information or enquiries. This proactive approach to the provision of information would be well received by the project team.

Research strategies can involve:

• obtaining existing health and safety files;
• assessing old drawings;
• obtaining local authority records;

- contacting previous clients, building managers, etc.;
- commissioning a conditional survey to act as an initial strategy;
- knowledge of the proposed construction project to assess the interfaces with existing structures, services and the use of the existing environment.

An appreciation and understanding of the construction processes and other CDM duty holders' responsibilities can also be very helpful in deciding how far to go in providing relevant information.

Historically this information-gathering exercise has been carried out by the designers, but ensuring its provision has been clearly defined in the CDM Regulations as the client's duty. This does not mean that the client cannot make project arrangements for the project manager perhaps to undertake on behalf of the client. However, liability will remain with the client.

Clients must bear in mind that this duty has many business risk issues. The cost in time and money of not identifying problems that can be considered in the initial design stage and managed during the construction phase can be significant.

Ensuring no work begins on site until a construction phase health and safety plan has been suitably developed

Again this is an important client duty in that absolutely no work can legally start on site until the principal contractor's health and safety plan has been deemed suitably developed.

> Every client shall ensure, so far as is reasonably practicable, that the construction phase of any project does not start unless a health and safety plan complying with regulation 15(4) has been prepared in respect of that project.
>
> CDM regulation 10

The planning supervisor will be available to advise the client on the initial contents of the construction phase health and safety plan if required. Indeed it is often the planning supervisor who actually deems the plan suitably developed for work to begin on site. However, note that the duty and consequently the liability will always remain the client's.

To emphasise the significance of this duty, CDM regulation 10 is one of only two regulations that have no exclusion from civil liabilities. Out of interest, the other is the principal contractor's duty for the security of the site.

The value of this 'procedural gateway' in the regulations is best demonstrated on the smaller projects where contractors traditionally do not have the same resources in terms of health and safety.

It could also be fair to say that complacency and informality can sometimes lead to a scheme physically beginning on site prior to any client declaration.

The Health and Safety Executive would not look lightly on any project where enabling works were being undertaken or site setup was under way before the client had formally deemed the construction phase health and safety plan suitably developed for work to start. The enforcement authority has carried out numerous successful prosecutions of clients who have failed to assess the health and safety plan prior to construction work starting on site.

All clients would be well advised to instigate a procedure for ensuring this duty is highlighted. Provision should also be made in terms of sufficient time if the plan is not actually suitably developed or falls short of the standard expected.

Availability of the health and safety file

The health and safety file contains information about a project that may be of use for the health and safety management of any future work on the structure. It can be considered as the operations and maintenance manual for hazards. The planning supervisor will ensure that the file is handed over to the client ideally as the project comes to its conclusion or at worst very soon afterwards. In some cases the file may even need to be handed over prior to completion, if areas of a structure are occupied for example.

> Every client shall take such steps as it is reasonable for a person in his position to take to ensure that the information in any health and safety file that has been delivered to him is kept available for inspection by any person who may need information in the file. . . .
>
> Extract from CDM regulation 12(1)

The professional construction client appreciates the benefits of a thorough and accurate health and safety file. Over and above the legal duty to make the file available for inspection there are obvious financial benefits to be made by informing maintenance operators and others of significant risks associated with future construction work, maintenance operations, cleaning work and even eventual demolition or dismantling of a structure.

The design and construction of a structure may account only for a small percentage of its overall cost, as illustrated in Figure 2.1. Building operators or facilities managers generally embrace the file, as its contents help reduce lifetime costs by making relevant health and safety information readily available to anyone needing to work on the structure.

From a client's business risk perspective too, the health and safety file is a valuable asset. Costs will be incurred throughout the life of a structure, and a live, assessable health and safety file will undoubtedly assist in reducing whole-life costs.

Traditionally, however, the development of the health and safety file has been left all too late in the project, and the planning supervisor, who is responsible

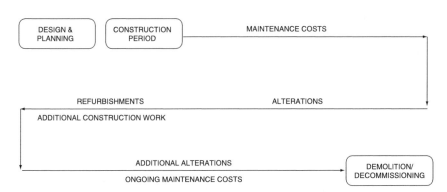

Figure 2.1 The life cycle of a structure.

for ensuring that the file is indeed handed over to the client, is left with a very difficult task. If little attention is paid to the health and safety file, much of the information required may be very difficult to acquire.

The client needs to discuss the health and safety file at an early stage with the planning supervisor, and consideration must be given to its structure, content and format. The emphasis in terms of format should be placed on the most appropriate, user-friendly approach for both accessing the information and updating. Formats can range from a single lever arch file or a CD Rom to a sophisticated web-based solution. Consideration must be given to the security and integrity of the file, whether a paper copy or an electronic copy. The availability of the information for inspection must remain as the key consideration as indeed the regulations demand.

Clients who have many projects where they also maintain the buildings would benefit from discussing the health and safety management issues with the building managers and health and safety representatives. Information in the form of accident statistics and experiences from contractors undertaking renovations or component replacement would help focus the information required for the file for all future activities. Even something as obscure as the weight of a glass curtain walling panel or type of paint finish used on a material can contribute significantly to health and safety in a practical way and add value to the client's organisation.

Also noteworthy is that, if and when a client hands a structure over to another party, for example if a building is sold, the heath and safety file must be provided to the new owner, ensuring the new owner is aware of its 'nature and purpose'.

The client's agent

The client will be informed of their statutory CDM duties, ideally by the designer before any work begins, and essentially must decide whether or not they are sufficiently competent and adequately resourced to carry out those duties.

If the client decides that they are not especially comfortable with the role for whatever reason, they are able to employ a separate entity who will assume full responsibility, but solely for the client's CDM duties.

> A client may appoint an agent or another client to act as the only client in respect of a project. . . .
>
> Extract from CDM regulation 4(1)

Legally, however, the client must ensure that anyone they engage to assume their responsibilities is suitably competent to carry out the client duties in respect of the CDM Regulations.

For the position of client's agent to be recognised by the Health and Safety Executive, a formal declaration must be made to them in writing.

The client's agent could be an existing member of the project team, for example the planning supervisor, contract administrator or project manager, or an external appointment.

There are generally two schools of thought when discussing the effectiveness of the client's agent. The first and more constructive is the idea that a client lacking in confidence in terms of their duties can engage a competent CDM professional to assume the role and add value to the project by fully embracing the regulations. Secondly, and more deviously, is the notion that clients may ignore all their statutory CDM duties and yet still control the project in all other ways. Clients would effectively not have to worry about their health and safety obligations but would continue to dictate in terms of project timescales, finance and other key drivers. It is argued by this lobby that the client can, if they so desire, disregard health and safety entirely and ride roughshod over the client's agent (who is legally charged with complying with the CDM Regulations).

It is this latter scepticism that appears to be winning the battle in terms of the future of the client's CDM agent. The proposed changes to CDM may suggest the complete removal of this option. This will ensure that the client cannot simply sidestep their significant health and safety management responsibilities by asking someone else to undertake the role.

Some frequently asked client questions

Q. *Am I competent to comply with my CDM duties as a client?*

A. Generally speaking, if one has to ask the question there could be a problem. There is no CDM duty for the client to be competent but there is a duty for employers to have health and safety arrangements in place in respect of their health and safety policy and to demonstrate compliance with regulation 5 of the Management of Health and Safety at Work Regulations 1999.

When it comes to the definition of competence as a client, if one knows the respective CDM duties and how to comply with them in an appropriate manner based on the size, complexity and risk profile of the project then one could be considered competent.

However, at the start of any project the designer is required to 'take reasonable steps' to inform the client of their duties, with the requirement in the Approved Code of Practice to provide a more comprehensive interpretation for less experienced clients. This can involve emphasising the client's role and good health and safety management and making early appointments of the other duty holders.

In making the judgements required to comply, the client should bear in mind that the CDM Regulations have been designed to assist the client where potentially difficult judgements are to be made. In respect of assessing the competence of designers and contractors and in respect of assessing the suitability of the construction phase health and safety plan, the client can call on the planning supervisor to offer advice, and the planning supervisor is duty bound to assist. Best practice should dictate that these service details are defined at the start of a project for clarity and resourcing.

The regulations also have the legal facility for the client to appoint an agent to act on their behalf. So in summary, depending on the advice from the designer on the CDM Regulations, one can make use of the planning supervisor's skills to facilitate compliance, and where the client does not feel competent or resourced to comply with the regulations an agent can be appointed.

Q. *As a client I have many projects which are borderline full CDM projects. Are there any guidelines or is there any advice I could get?*

A. The regulations and Approved Code of Practice provide guidance on the application and scope of the regulations. Borderline cases where it must be determined whether the full extent of the CDM Regulations applies or not are decided on the five or more persons rule, notification and whether demolishing or dismantling is involved.

Problems also potentially arise when, owing to a change, the project needs to be notified to the HSE, i.e. the programme extends significantly beyond 30 working days or 500 person working days. Other problems occur in defining the interpretation of a structure and in assessing the qualification of demolition and dismantling. If a highway authority removes one kerbstone from the highway, which is a structure, then does the full extent of the CDM Regulations apply? Another example could involve the cutting and knocking of a new doorway through an existing wall. If the wall is structural and load bearing, the full extent of the CDM Regulations applies; but if the wall is a partition wall and not load bearing is it a part of the structure?

Demolition and dismantling

CDM applies to all demolition or dismantling work, whether or not the work has to be notified to the Health and Safety Executive (HSE). Demolition and dismantling includes the deliberate pulling down, destruction or taking apart of a structure, or a substantial part of a structure. It includes dismantling for re-erection or re-use. Demolition does not include operations such as making openings for doors, windows or services or removing non-structural elements such as cladding, roof tiles or scaffolding.

Extract from Approved Code of Practice

When assessing the application of the CDM Regulations to a project that is potentially just under 30 days, the Approved Code of Practice provides the following guidance:

Details of some projects which CDM applies to must be notified to HSE. This is the case where construction work is expected to:

- last more than 30 days; or
- involve more than 500 person days.

Any day on which work takes place counts towards the period of construction work, including weekends and holidays. The total 'person days' is the total number of shifts worked by everybody involved in the project, including supervisors. In borderline cases, where you are unsure how long the work will take, it is best to notify HSE.

Where a small project, which was not notifiable, requires a short extension or small increase in the number of people, you do not need to notify HSE of the change. However, construction work may significantly overrun or the scope may change so that it becomes notifiable. If this happens you must notify HSE as soon as you can.

Extract from Approved Code of Practice

The HSE's Approved Code of Practice was designed to provide guidance to the industry on the approach one should take to comply. The ACoP provides examples of how to resolve problems in assessing the extent of the CDM Regulations. However, one should never forget the objectives of health and safety and the respective regulations in that, whether or not the CDM Regulations apply, hazards must still be managed and the client or organisation involved must still have suitable health and safety arrangements. Under the Management of Health and Safety at Work Regulations 1999 (MHSW) most of the actions within the CDM Regulations are covered anyway. The client has to provide information at the start of the project, risk assessment must be undertaken for hazardous activities, safety plans in the form of method statements must be produced and assessed, and safety information that is required for future works must be provided. Also people must be trained to do the work safely. So if the CDM Regulations do not apply you generally have to go through the same procedures using the MHSW Regulations 1999.

In summary, the ACoP provides the majority of the answers. If then you still feel you need advice, contact your safety adviser or the Health and Safety Executive.

Q. *How far do I go in providing information on the project, the structure and the environment to the planning supervisors and designers?*

A. The regulations state that:

> The information required to be provided . . . is information which is relevant to the functions of the planning supervisor under these Regulations and which the client has or could ascertain by making enquiries which it is reasonable for a person in his position to make.
>
> CDM regulation 11(2)

The Approved Code of Practice tells us that the level and quality of information must ensure that significant risks are predictable and effectively managed. There is a selection of examples of type of information earlier in Section 2 under the heading 'Providing relevant health and safety information to the team', but this is by no means exhaustive. In terms of how far one takes the term 'reasonable enquiries', a balanced approach is necessary based on the level of risk. However, any decision to omit information which consequently led to a serious injury or worse would need to be defended.

Q. *When do I need to appoint a planning supervisor?*

A. The CDM Regulations make this very clear:

> The planning supervisor shall be appointed as soon as is practicable after the client has such information about the project and the construction work involved in it as will enable him to comply with the requirements imposed on him by regulations 8(1) and 9(1).
>
> CDM regulation 6(3)

It would be prudent to engage the planning supervisor when the design team are at the concept stage, as this is when they can arguably make their largest contribution to health and safety management.

The late appointment of the planning supervisor is a widespread industry failing and has undoubtedly led to numerous lost opportunities at the early design stages.

Q. *How do I appoint a qualified planning supervisor?*

A. It is important to understand that there is no such entity as a 'qualified' planning supervisor. Individuals or organisations offering the service must be competent and adequately resourced for the specific project in question.

Membership of any planning supervisory institute or association is not therefore a qualification and actually is no guarantee of competence.

Q. *How do I find a planning supervisor?*

A. Many clients, especially the less experienced, will take guidance from the designer. Many designers offer the service themselves or will know of individuals or organisations that can offer the service. Remember, however, that it is the client who is legally charged with ensuring that the appointee is both competent and adequately resourced for the project and not the designer. If the client appoints a planning supervisor based on a recommendation, it may not satisfy the CDM Regulations.

Q. *As the client am I responsible for health and safety standards when construction work begins on site?*

A. The client has no statutory duty under the CDM Regulations to monitor on-site health and safety performance. The principal contractor is responsible for health and safety management during the construction phase.

However, this is not to say that the client should ignore on-site health and safety standards. Section 2 of the Health and Safety at Work etc. Act 1974 (HSWA) states that:

> It shall be the duty of every employer to ensure, so far as is reasonably practicable, the health, safety and welfare at work of all his employees.

Clients may therefore need to consider how any construction work may affect their employees. An example may be the impact an office extension project will have on existing employees.

Similarly, section 3 of HSWA defines duties owed to others who may be affected by the employer's undertakings to ensure that they are not put at risk. Examples include the general public, visitors and other non-employees.

Section 4 of HSWA also imposes health and safety duties on controllers of non-domestic premises. They need to ensure, as far as is reasonably practicable, that the premises, the means of access and exit, and any plant or substances used are safe and without risks to health.

CDM clients therefore may well need to make provisions to monitor health and safety arrangements to satisfy the above duties imposed by HSWA. Many professional clients will also monitor the performance of the principal contractor as a means of evaluating competence for similar-type work that may be required in the future.

Q. *How do I know whether a principal contractor's construction phase health and safety plan is good enough for work to begin?*

A. The ACoP suggests a list of minimum headings that should be in place from the start of the construction work, namely:

(a) general information about the project, including a brief description and details of the programme;
(b) specific procedures and arrangements for the early work;
(c) general procedures and arrangements which apply to the whole construction phase, including those for the management and monitoring of health and safety;
(d) welfare arrangements;
(e) emergency procedures; and
(f) arrangements for communication.

If a client's CDM agent has not been appointed and the client is not competent to assess the quality of the construction phase health and safety plan, then the planning supervisor will be available to advise on the plan's initial content. However, occasionally planning supervisors may feel obliged not to slow down a project by suggesting the plan is not suitably developed for work to start. A client must not make the assumption that the plan will be good enough and must allow the planning supervisor licence to properly assess its content and report accordingly. It is in the client's interest more than anyone's to ensure that the construction phase health and safety plan is appropriate and suitably developed before work starts.

Chapter 3

CDM designers

Introduction

This section has been designed to provide the designers with the knowledge and day-to-day skills, the solutions to common problems, and the ability to assess their own competence and development and where possible to see the opportunities to add value to their service and the project as a whole.

The statutory duties imposed on designers by the CDM Regulations were designed to set health and safety standards at the earliest stages of design during a project and to take advantage of the opportunities that pre-planning and design professionals have to address any potential health and safety site and future maintenance problems proactively.

Project safety standards are potentially set before the start of the design phase by the designer complying with their duty to advise the client of their specific duties to comply with the CDM Regulations. Design opportunities exist to proactively address any health and safety issues that are managed by designers through identifying significant hazards and taking appropriate design action to avoid or reduce the risks.

Significantly too, the designer's duties are the only duties to apply to all construction projects where the HSE are the enforcing authority, which is the majority. This suggests that even on the smallest of projects a designer can contribute to avoiding and reducing risks.

Designers' opportunities to contribute to the management of health and safety on a project extend to the selection of building materials, spatial design, obtaining site information and informing contractors about difficult-to-manage hazards. The fundamental objective of the designer's duties is to design out significant hazards. The design management skill of coordination and cooperation with other designers and suppliers to address a problem as a team also adds significant value to any project.

In simple terms the designers on any project must have the design and health and safety skills and knowledge to proactively identify significant health and safety problems (hazards) and through their design service take action to ensure the structure can be constructed, and maintained, safely.

It is true to say that the design fraternity and the CDM Regulations have not had the closest of relationships since they came together in 1995. It was widely thought that many design practices did not perhaps fully appreciate the contribution they could make to construction health and safety. Whether designers were mis-sold the principles of CDM or whether they misunderstood what was required of them, the fact remains that numerous practices did not go far enough in embracing this legislation. However, evidence was necessary to prove that designers were failing to address the occupational health and safety implications of their designs.

Consequently a Health and Safety Executive campaign was undertaken in 2003 to assess the levels of competence and compliance throughout numerous design practices in the north of England and Scotland. It highlighted some major shortcomings:

- About a third of the designers demonstrated little or no understanding of their responsibilities, and only 8 per cent claimed to have received training in CDM.
- Many of the design risk assessments were of poor quality and added little if anything to the safety of the construction process.
- Designers were often abdicating their responsibility to reduce risk in relation to work at height by leaving it to the principal contractor, without first considering how they could change the design in a way which would make it safer to build, clean or maintain.
- Contractors were struggling to control risk which could easily have been eliminated or considerably reduced by good design.

Source of data: HSE Designer Initiative 2003

Following the initial visits to design practices the HSE saw some improvements but based on the total number of prosecutions it is evident that the legislation and enforcers are struggling to have an impact. More evidence for this is contained in HSE Research Report 218, which was based on a review of 73 accidents and incidents. The conclusions drawn identified that 65 per cent of incidents could support a designer prosecution under the CDM Regulations.

Who are CDM designers?

With a clear understanding that designers have legal duties and in complying with them have real opportunities to contribute to the health and safety on projects and throughout the life cycle of a structure, one important question should be 'Who are designers?' The influences on and contributions to a design are not solely under the control of the traditional architects and engineers. To clarify this issue for organisations and clients setting standards and service arrangements on projects, the regulations have a broad definition of a 'designer':

Designers are those who have a trade or a business which involves them in:

- preparing designs for construction work including variations – this includes preparing drawings, design details, specifications, bills of quantities and the specification of articles and substances, as well as all the related analysis, calculations, and preparatory work; or
- arranging for their employees or other people under their control to prepare designs relating to a structure or part of a structure.

CDM Approved Code of Practice

This definition can be more accurately broken down to include a wide variety of designers with a set of CDM duties and the potential to manage and contribute to health and safety. Examples of who is included are listed below:

- architects, civil and structural engineers, building surveyors, landscape architects, and design practices (of whatever discipline) contributing to, or having overall responsibility for, any part of the design, e.g. drainage engineers designing the drainage for a new development;
- anyone who specifies or alters a design, or who specifies the use of a particular method of work or material, e.g. a quantity surveyor who insists on specific material or a client who stipulates a particular layout for a new production building;
- building service designers, engineering practices or others designing fixed plant which people can fall more than 2 metres from (this includes ventilation and electrical systems), e.g. a specialist provider of permanent fire extinguishing installations;
- those purchasing materials where the choice has been left open, e.g. people purchasing building blocks and so deciding the weights that bricklayers must handle;
- contractors carrying out design work as part of their contribution to a project, e.g. an engineering contractor providing design, procurement and construction management services;
- temporary works engineers, including those designing formwork, falsework, scaffolding, and sheet piling;
- interior designers, including shop-fitters who also develop the design;
- anyone specifying or designing how demolition, dismantling work, structural alteration, or formation of openings, is to be carried out; and
- heritage organisations who specify how work is to be done in detail, e.g. providing detailed requirements to stabilise existing structures.

CDM Approved Code of Practice

As a rule any official, for example a building control officer, or planning officer who is appointed and authorised under a specific Act or regulation, is not a CDM designer irrespective of their influence over the design and its risk profile. How the designer addresses professional concerns on projects regarding these issues is covered later in this section, but the designer is responsible for managing the health and safety aspects of that design change brought about by the official.

Designers' duties explained

Introduction

Fundamentally there are currently four designer's duties and one other general duty that specifically applies to designers when they engage the services of other designers. The duties are to:

- take responsible steps to inform the client of their duties under the CDM Regulations;
- give adequate regard to the hierarchy of risk control when carrying out design work;
- ensure design includes adequate information about health and safety;
- cooperate with the planning supervisor and other designers so they can comply with their duties;
- ensure, when arranging for any designer(s) to prepare a design, that they are competent and adequately resourced for health and safety.

We will now examine in detail how the designer should comply with these duties and provide guidance on compliance and also adding value.

Informing the client of their duties

This regulation is to be executed by the designer before any design work starts and has the practical, legal and potential added value of protecting the client's interests with the associated business risk benefits. The client will be informed that they must initially provide, to the planning supervisor, information on the any existing structure(s), the project, and environment in relation to health and safety. They will also be told that a planning supervisor and principal contractor must be appointed and they and any designers must be competent and adequately resourced to address the health and safety issues to be involved. Finally advice will be given on not allowing construction work to start until a construction safety plan has been suitably developed by the principal contractor. All these client duties add value in providing opportunities to manage health and safety from the onset and significantly during future maintenance with the management and availability of the health and safety file.

> Except where a design is prepared in-house, no employer shall cause or permit any employee of his to prepare for him, and no self-employed person shall prepare, a design in respect of any project unless he has taken reasonable steps to ensure that the client for that project is aware of the duties to which the client is subject by virtue of these Regulations and of

any practical guidance issued from time to time by the Commission with respect to the requirements of these Regulations.

CDM regulation 13(1)

The regulation, in using terminology like 'taken reasonable steps to ensure', is not absolute and requires a judgement by the designer that must be based on their assessment of the client's experience and technical knowledge, i.e. competence. This regulation has a legal gateway obviously designed to promote compliance in not permitting the preparation of a design unless the client is aware of their duties.

The CDM Approved Code of Practice (ACoP) provides guidance and advice that as a minimum the client should be made aware of, in particular the first three sections of the ACoP covering managing construction projects, application, interpretation and notification of these regulations and the client's duties. Also as a minimum standard the ACoP advises the client to have the HSC leaflet 'Having construction work done? Duties of clients under the Construction (Design and Management) Regulations 1994'.

More advanced methods of adding value can include arranging presentations at the start of a project which can involve the Health and Safety Executive or the contribution of an experienced consultant. A more direct way of adding value that is over and above the regulations could be to provide the client with some of the CDM administrative tools necessary to guide the client through the process and where a specific task is required to provide the necessary checklist or pro forma. The tools that would assist a client could be:

- a client's CDM project checklist;
- a hazard and information checklist to identify information at the start of the project;
- a construction phase health and safety plan approval checklist;
- a competence and resources questionnaire, interview agenda and appraisal pack.

There is no legal duty to provide these tools but they provide opportunities to have a more engaged client and proactive approach.

One other issue a designer must consider in advising the client is whether they need the support of a CDM client's agent. Where it is evident that the client is not competent or resourced to undertake or comply with their duties the recommendation to employ an agent is an option to consider. Another option is to contract the services of the proposed project team and allocate responsibility through a clear CDM service level agreement or set of project arrangements. This could incur additional costs and would not negate the ultimate responsibility of the client to ensure the duties are addressed. The planning supervisor currently has duties to provide advice if requested on the competence and resources of designers and contractors and also to advise on the suitability of the construction phase health and safety plan, which are two duties that potentially require a higher level of competence.

Case study 1: A design practice has introduced an in-house procedure to issue the HSE leaflet ('Having construction work done?') with their letter of appointment. This is supplemented by the address of their website, which has a specific client's CDM page providing additional advice and information.

In summary, this designer duty provides an early and invaluable opportunity to educate the client at the start of any size of project and takes advantage of the early client–designer interface. From the client's business risk and investment perspective and from a hazard management perspective, it is a very important regulation with potentially significant implications if the designer does not carry out the duty effectively. The CDM competent designer would benefit from adding value in providing the client with the necessary tools to demonstrate their commitment to CDM and the project. It could pay dividends if something went wrong later on and the project team were subjected to an accident investigation.

Designing safely

Designing safely, following on from ensuring the client is aware of their CDM duties, is the second fundamental way a designer can make a significant health and safety contribution. The regulations impose a duty to design with adequate regard to health and safety, which is qualified by more specific regulations to avoid foreseeable risks, combat risks at source and give priority measures to protect all persons at work including persons affected by the work, i.e. the public or existing building users.

Simple examples of these duties range from spatial design options to locate a building, wall, etc. in such a position as to avoid an interface with existing asbestos or to select a concrete block that weighs less than 20 kilograms to significantly reduce the risk of a manual handling injury.

The skill of the competent CDM designer is how to manage and integrate this management process into the design service so the designer and design team can take advantage of the opportunities to design safely.

We will now go on to cover the procedures that make up the hazard management process. This is illustrated in simplified form in Figure 3.1.

Case study 2: A designer has commissioned and specified a pre-cast bespoke staircase system that would be difficult to install after the main structure has been completed. Because it is potentially difficult to manage on site, the designer has made a clear note on the drawing that the stairs will have to be installed at a critical time during construction.

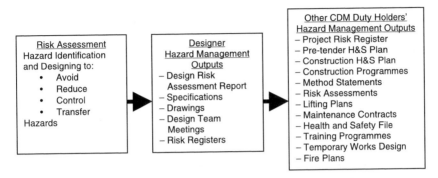

Figure 3.1 Design hazard management.

Hazards

A hazard has a life cycle, and a designer's knowledge of hazards and their associated relationship with the construction and maintenance processes and environmental issues is essential. Without this knowledge and experience of construction a designer will find it impossible to comply and make a tangible health and safety contribution.

> *Definition of hazard:* Something with the potential to cause harm

This section will cover the difficult questions of what hazards designers have to identify and consider as a part of their design and how far they have to go in respect of designing them out or reducing the risks. It is assumed that a competent designer has buildability knowledge and can foresee potential design conflicts, obvious health and safety problems and difficult-to-manage issues on site, but what does the inexperienced designer do? This section will also address the questions concerning a variety of options to implement a hazard management strategy and address the subject of risk assessment.

The issues concerning the extent of liability a designer has in respect of design elements not under their control and who coordinates health and safety during design will also be covered. A common misconception is that the planning supervisor is the coordinator and responsible for managing the design's health and safety element.

Design risk assessment

The whole issue of design risk assessment has been confusing for the design fraternity since the inception of the CDM Regulations. The regulations do not ask for designers to assess risks but to design with adequate regard to health and safety. The CDM ACoP also states that in most cases it is sufficient to use existing experience and published guidance; there is no need for sophisticated risk analysis techniques. However, the industry has interpreted this duty as risk assessment. Recently, though, the Health and Safety Executive and other industry organisations have promoted alternatives to risk assessment. These alternatives, however, encourage very similar concepts to address the duties for designing safely. It is the author's belief that the methodology and tool to be deployed for compliance and safe designing should be based on the factors associated with the designer's competence and experience and the risk profile of the project.

The fundamentals of designing safely are based on the identification of hazards and using the designer's skills to take action using a hierarchical response, as illustrated in Figure 3.1, but to do this the designer must have a knowledge of hazards and the construction methods or materials from which they derive. Risk assessment is not a process of recording residual risks at various design stages but should be considered as a design mindset that integrates health and safety into the design development and intuitive decision-making processes. The risk assessment report or pro forma is only a summary of design management actions and should be considered a real-time report for communicating actions to other designers, the planning supervisor and the principal contractor. Depending on one's policy arrangements or approach, design mitigation can also be recorded for audit purposes perhaps or even sharing good practice. As intended with a pro forma, the form can help the user in going through a process and act as an aide-memoire, depending on design. More modern approaches aimed at assisting designers consist of IT system databases that prompt the designer to consider issues of the design and record them as necessary.

Some designers do not see the need for a pro forma as a hazard management tool. They are confident they can use existing means of managing design information to record, communicate and project-manage the health and safety aspects of the design. These obviously include the drawings, specifications, minutes of meetings, etc. The risk assessment pro forma can in its simplest form be used to record the residual risks only that the other designers, construction contractors and future maintenance contractors will have to consider and manage.

Based on the objectives of the exercise the strategy of risk assessment, i.e. how and what you assess, must be proportionate to the risk profile and not the size or price of a project. A large housing project could have a low-risk profile because of the location and simplicity of design but could be worth millions of pounds. Alternatively, a low-cost extension could have many environmental, structural, logistical/interface and building hazards which make the designer's contributions more extensive when designing with adequate regard to health and safety.

The CDM Regulations offer designers the following:

Every designer shall –

(a) ensure that any design he prepares and which he is aware will be used for the purposes of construction work includes among the design considerations adequate regard to the need –

   (i) to avoid foreseeable risk to the health and safety of any person at work carrying out construction work or cleaning work in or on the structure at any time, or of any person who may be affected by the work of such a person at work,

   (ii) to combat at source risk to the health and safety of any person at work carrying out construction work or cleaning work in or on the structure at any time, or of any person who may be affected by the work of such a person at work, and

   (iii) to give priority to measures which will protect all persons at work who may carry out construction work or cleaning work at any time and all persons who may be affected by the work of such persons at work over measures which only protect each person carrying out such work.

Extract from CDM regulation 13(2)

The regulations go on to qualify the above in that they:

> shall require the design to include only matters referred to therein to the extent that it is reasonable to expect the designer to address them at the time the design is prepared and to the extent that it is otherwise reasonably practicable to do so.

Extract from CDM regulation 13(3)

So designers are provided with a defence that appreciates they cannot design safely for elements of the design that do not have a clear brief or where insufficient information exists on other aspects of the design, e.g. mechanical and electrical plant size and location. The defence also defines the extent to which the health and safety elements of design can be weighed against the other design and project issues such as cost, aesthetics, fitness for purpose and environmental issues.

The day-to-day application of the regulations requires a strategy to focus the designer's attention. The strategy must have the ability to identify the potential problems relating to the environment and construction elements of the design. One common starting approach is hazard identification that can be part of a project risk management process or one of the defined first steps of a risk assessment strategy.

The designer could not practically and is not legally required to identify all hazards, only foreseeable ones. The identification of hazards obviously requires knowledge of hazards and what hazards are significant and cause fatalities, major injuries and occupational and public health problems.

The pro forma in Figure 3.2 is designed to provide the designer with a strategy to examine the main construction and maintenance issues on the project and identify which hazards are applicable and require consideration within the design.

The mindset or more formal strategy that can be prompted by an aide-memoire or pro forma should identify the main assumed activities associated with the project and in doing so provide a platform or opportunity to consider what significant hazards are associated with the activity. For example, a knowledge or assumption that deep excavations will be needed provides an opportunity to consider the associated hazards with the construction activities and consequently what design action can be taken to design with adequate regard to heath and safety. This type of knowledge regarding buildability is essential and is one of the key characteristics of a competent designer.

The exercises below give a simple illustration of mindsets to provide guidance on a strategy of how to identify hazards, which can be started as an initial brainstorming session and also used at the various stages of design as a review agenda.

Exercise: Strategy and mindsets from the perspective of the activities
View Figure 3.3 and consider the activities and interfaces associated with the:

1   surrounding environment;
2   site setup and management;
3   building/construction;
4   maintenance.

Exercise: Strategy and mindsets from the perspective of the hazards
View Figure 3.4 and consider the activities and interfaces associated with the:

1   surrounding environment;
2   site setup and management;
3   building/construction;
4   maintenance.

Thus far we have explored activities associated with our design and the hazards that may be associated with those activities. This is summarised in Figure 3.5.

As we should feel confident about how we identify hazards and what is a significant risk we will now concentrate on the design actions and prioritising those actions. The design action outputs from a risk assessment can be classified for clarity and are illustrated in Figure 3.6.

If the simple principles of design risk assessment are identifying a problem and then taking design action to make the construction and maintenance safer where it is practicable then we are now focusing on the design action order. We call this order or hierarchy the 'principles of prevention', and it is derived from

Project......3 Story School Extension/ St Davids................**Design Stage**.........Scheme...................**Date**......15th April 2006..............
Designer/Organisation......A Designer & Co...

Design Area & Activity Insert reference No in appropriate box.	**Hazards** (NB- only select hazards with significant risks)																			
	Hazardous Substances	Confined Spaces	Fall From Height	Falling Objects	Site Plant Vehicles	Collapsing Structure	Manual Handling	Moving Objects	Electricity Services	Gas Services	Fire Explosion	Noise & Vibration	Cuts & Abrasions	Asbestos & MM Minerals	Fire Means of Escape	Highway Traffic	Adverse Weather	Access for Cleaning	Access for Maintenance	Designer/ Company Responsible
Site Set Up																				
Site Workface Access																				
Deliveries																				
Welfare Location																				
Pedestrian Routes																				
Temp Services																				
Car Parking																				
Materials Storage																				
Hoarding																				
Construction &																				
Excavating																				
Piling																				
Strip Foundations																				
Slab																				
Superstructure																				
Windows & Atria																				
Maintenance Work																				
Window Cleaning																				
Component Replacement																				
Light Fittings																				

Figure 3.2 Hazard identification pro forma.

3. Building Activities
Substructure
Ground Clearance
Excavations
Foundations
Superstructure
Steel Frame Erection
Pre-Cast Concrete
Lift Shafts
Block/Brick Masonry Work
Lifting Installing Beams
Roof
Lifting Truss sections
Membrane Coverings
Roof Cladding
Fitting Sky Lights
M&E
Installing Lifts
Electrical Installation
Mechanical Installation
Windows
Installing Doors
Installing Windows
Finishes
Rendering Plastering
Dry Lining

4. Maintenance Activities
Window Cleaning, Large Component Replacement, M&E Maintenance,
Refurbishment/New Internal Layout

Proposed New Building

Existing Building

Surrounding Environment

1. Logistical Activities/ Surrounding Environment
Planning Highway Deliveries
Off Site Storage of Materials
Temporary Highway Access

Extent of Site

2. Site Activities
Locating and Accessing Site Offices, Materials Compound &
Welfare, Location of Cranes, Traffic Management Routes,
Installing Scaffolding, Car Parking, Controlling Unauthorised
Access, Waste Management, Temporary Services

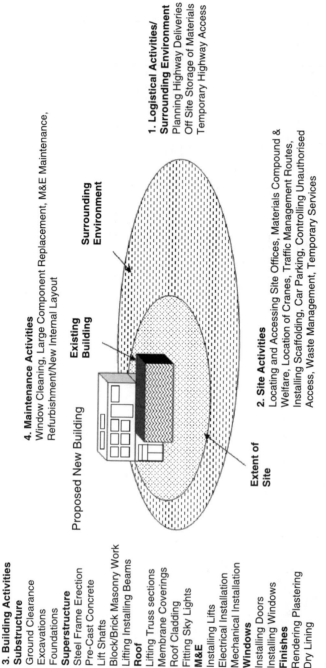

Figure 3.3 Strategy and mindsets from the perspective of the activities.

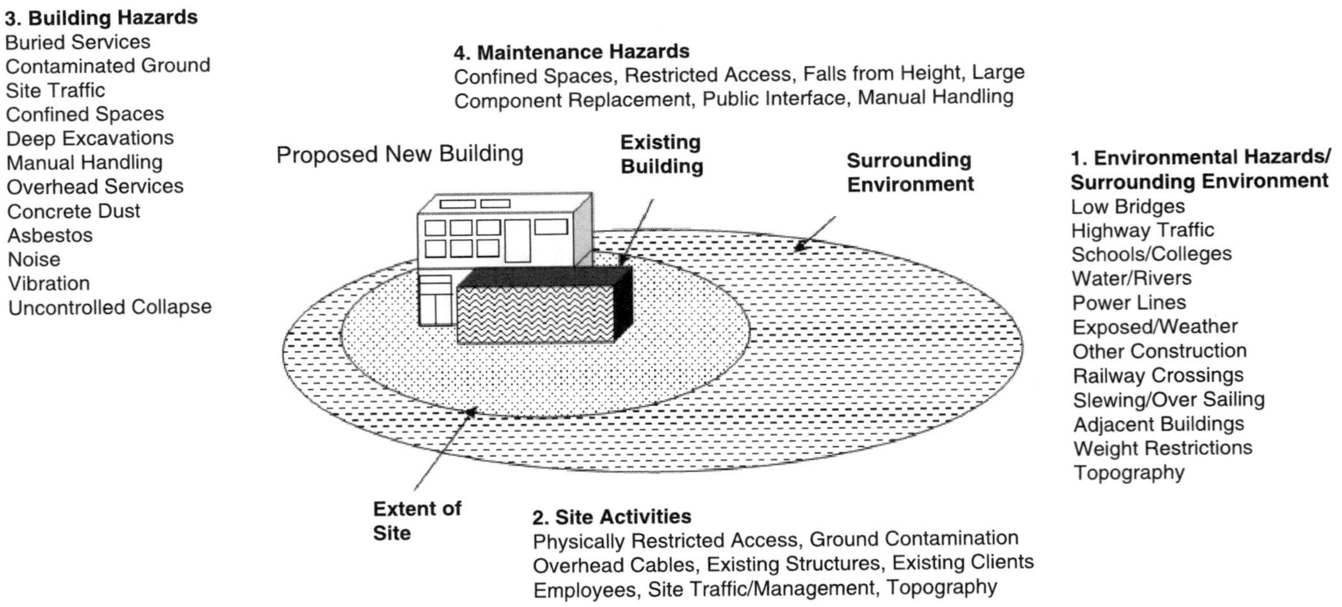

3. Building Hazards
Buried Services
Contaminated Ground
Site Traffic
Confined Spaces
Deep Excavations
Manual Handling
Overhead Services
Concrete Dust
Asbestos
Noise
Vibration
Uncontrolled Collapse

4. Maintenance Hazards
Confined Spaces, Restricted Access, Falls from Height, Large
Component Replacement, Public Interface, Manual Handling

**1. Environmental Hazards/
Surrounding Environment**
Low Bridges
Highway Traffic
Schools/Colleges
Water/Rivers
Power Lines
Exposed/Weather
Other Construction
Railway Crossings
Slewing/Over Sailing
Adjacent Buildings
Weight Restrictions
Topography

Surrounding
Environment

Existing
Building

Proposed New Building

Extent of
Site

2. Site Activities
Physically Restricted Access, Ground Contamination
Overhead Cables, Existing Structures, Existing Clients
Employees, Site Traffic/Management, Topography

Figure 3.4 Strategy and mindsets from the perspective of the hazards.

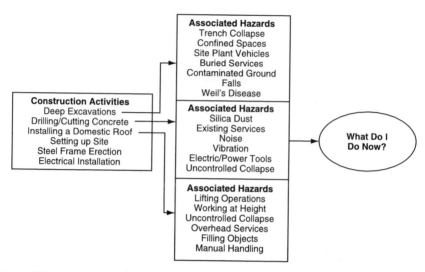

Figure 3.5 Activity and hazard relationship.

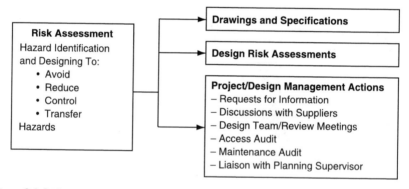

Figure 3.6 Risk assessment outputs.

Schedule 1 of the Management of Health and Safety at Work Regulations 1999 (MHSW). This is illustrated as follows:

> In deciding upon which preventive and protective measures to take, employers and self-employed people should apply the following principles of prevention:
>
> (a) if possible avoid a risk altogether; e.g. could the work be done in a different way taking care not to introduce new hazards;
> (b) evaluate risks that cannot be avoided by carrying out a risk assessment;
> (c) combat risks at source, rather than take palliative measures. So, if the

steps are slippery treating or replacing them is better than displaying a warning sign;

(d) adapt work to the requirements of the individual (consulting those who will be affected when designing workplaces, selecting work and personal protective equipment and drawing up working and safety procedures and methods of production). Aim to alleviate monotonous work and paced working at a predetermined rate, and increase the control individuals have over work they are responsible for;

(e) take advantage of technological and technical progress, which often offers opportunities for improving working methods and making them safer;

(f) implement risk prevention measures to form part of a coherent policy and approach. This will progressively reduce those risks that cannot be prevented or avoided altogether; and will take account of the way work is organised, the working conditions, the environment and any relevant social factors. Health and safety policy statements required under section 2(3) of the HSW Act should be prepared and applied by reference to these principles;

(g) **give priority to those measures which protect the whole workplace and everyone who works there, and so give the greatest benefit (i.e. give collective protective measures priority over individual measures);**

(h) ensure that workers, whether employees or self-employed, understand what they must do;

(i) the existence of a positive health and safety culture should exist within an organisation. That means the avoidance, prevention and reduction of risks at work must be accepted as part of the organisation's approach and attitude to all its activities. It should be recognised at all levels of the organisation from junior to senior management.

MHSW Approved Code of Practice regulation 4

(Note that the bold sections highlight the statements extrapolated into the CDM designer duties.)

For simplicity and practical application, more user-friendly interpretations are adopted to assist designers to prioritise the hazard management element of the design process. This complements the simplicity of hazard identification and provides designers with more potential to integrate the principles of risk assessment into their design service.

For a designer to be confident in designing with adequate regard to health and safety and to be able to demonstrate competence in risk assessment, a sound understanding of the control hierarchy is essential. The following paragraphs will explain, with practical examples, the control hierarchy applied to design. This has the objective of promoting a consideration in the designer of designing to avoid hazards altogether and, where this is not reasonably practicable, designing to reduce the level of risk or specifying a control measure.

Where this is not possible the designer will transfer the hazards by default, but not all hazards need to be brought to the attention of the contractor or other designers.

Avoidance

For a designer to avoid, through a design decision, a hazard requires a high-level but fundamental decision to be made. For environmental and site hazards it could be as simple as locating the structure away from a buried cable or a contaminated hot spot. This highlights the importance of the client's information under CDM regulation 11 to provide information on the site. Not interfacing with an asbestos-covered wall in favour of an alternative layout or selecting a pre-finished material to avoid working at height or a non-maintainable material to avoid a maintenance operation working at height is the simple tenet of risk elimination.

Avoiding a hazard, i.e. something with the potential to cause harm, by a design decision highlights one of the basic tenets and prime movers of the original directive and the UK's interpretation the CDM Regulations, which is to reduce the level of fatalities, accidents and ill health.

Certain design options that avoid hazards introduce others that need consideration. For example, specifying a prefabricated solution to a wall to avoid manually handling large concrete blocks introduces a crane to the site and the hazards associated with lifting operations. The judgement or assessment must be based on risk and it is on some sites possible to manage the lifting operations associated with a crane without difficulty, but, where the site interfaces with the public, overhead cables, etc., craning large components is difficult to manage

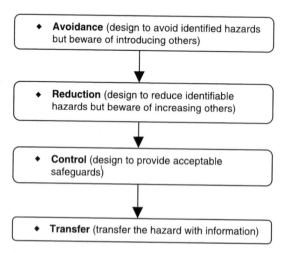

Figure 3.7 The control hierarchy.

safely and block-laying may be safer based on the designer's duty to take priority measures to protect the other site workers and others affected by the work.

Reduction

If a designer cannot avoid a risk altogether then they must consider reducing the risk to an acceptable level by assessment. Designers must, however, also beware of introducing other hazards that are difficult to manage for the contractor or maintenance operative.

To reduce the level of risk we obviously need to understand what constitutes a risk. A risk could be defined as the likelihood of harm and/or severity that the hazard could cause. So it is a product of two factors that qualify the hazard's potential to be worth considering. This is also helpful when deciding whether a hazard is significant, one of the principles of hazard identification.

A risk assessment is by its very nature subjective, so to have some mechanism to apply the principles is helpful. To achieve this we use the risk factor, which can be quantitative, e.g. high, medium or low, or based on a matrix or a product of two numerical factors for likelihood and severity as illustrated in Table 3.1.

The respective designers' relationship, designers' knowledge and assumptions of the construction and maintenance activities and subsequent associated hazards are the key elements to reducing risks. With an understanding of what is likely to cause harm, e.g. the difference between lifting a 36kg 140mm block as opposed to a 20kg 100mm block, the opportunity to reduce the likelihood of a back injury can be reduced. Similarly if the design requires the activity to be 1 metre above ground level then the severity of injury as a result of falling will be minimal as opposed to that from working 4 metres above ground level.

Based on these simple principles a designer can significantly reduce the risk of

Table 3.1 Using the risk factor

		SEVERITY		
LIKELIHOOD		L = 1	M = 2	H = 3
	L = 1	1	2	3
	M = 2	2	4	6
	H = 3	3	6	9

Severity × Likelihood = Risk factor

Severity =
Low (1) = Very minor injury or illness
Med. (2) = 3-day injury or short-term illness
High (3) = Fatality, major injury or long-term/terminal illness

Likelihood =
Low (1) = Slight or no chance of occurring
Med. (2) = Reasonably likely to occur
High (3) = Highly likely to occur

injury or ill health by reducing the number of hazardous operations. To reduce the depth of an excavation to 1 metre from 3 metres reduces the likelihood of trench collapse and, if the trench collapses, reduces the severity from a potential fatality to a minor accident. Other risks are also reduced, e.g. working in confined spaces, contaminated ground, falls, falling objects, buried services, etc.

Other examples include reducing the time workers spend at height by selecting low-maintenance products and reducing the number of penetrations through concrete walls for the M&E systems, which will reduce the operative's exposure to dust, vibration, noise, etc. This type of interface between design issues is where competent designers would coordinate matters to proactively 'design in' all service openings to avoid hazardous on-site activities.

Designing in the potential for a one-way traffic management route on site or alternative safer access to the site by creating a temporary access can also reduce risks significantly.

Control

Design actions to control hazards can vary from the specification of a man-safe system, an access gantry or a handrail on a leading edge to a specification to use specialist contractors to undertake specific work with inherent risks.

A control measure should be considered as a last resort but should not be considered bad design. The prolific use of man-safe systems on roofs that require minimal maintenance and the inclusion of simple eye bolts are a realistic improvement in high-risk maintenance safety.

Other control measures could include designing in features for safe working rather than specifying safety hardware. Examples of these include specifying the early installation of stairs and designing in hard landscaping or standings for access equipment.

All these controls are designed to reduce the risk to an acceptable and manageable level by the contractors or maintenance operatives. The terminology can confuse the designer in that a control measure reduces a risk and also transfers it to the contractor. As a guide, a reduction of risk should be considered as a design decision that reduces the number of operations, the exposure to a hazard or the likelihood of actual harm or ill health. A control is a designed-in feature that is either a safety product or a design feature.

Transfer

The issue of designing safely is focused on the hazard: the entity with the potential to cause harm. The post-risk assessment process is often referred to as hazard management. The hazard can be considered to have a life, and clients, designers and contractors have the task of managing the hazard at various stages within a project. The management of the hazard could be comparable to the management of a budget or any other project criteria.

The hazard can be transferred to another designer where they have more opportunity to reduce risk, by virtue of their design service and expertise, to a contractor for management on site or to the client for consideration as a project issue that requires an action.

Hazard management, like all aspects of design, requires a procedure which implements a set of objectives and rules to achieve the objectives and someone within the design team to manage and coordinate the safety elements. A lead consultant or designer who is managing and controlling the design element of the project is obviously best placed.

In transferring a hazard, a designer is managing it, but what hazards should be transferred and what information should be provided with the hazard are common concerns of the design fraternity. These issues will be addressed shortly, but basically the limitations to the designer's duties in providing information are based on only mentioning hazards that are unusual and difficult to manage and not mentioning hazards and issues one would expect a competent designer or contractor to be capable of managing effectively.

Note: Designers of commercial premises should, as best practice, consider the respective legislation and duties of the client or building user. The Workplace (Health, Safety and Welfare) Regulations 1992 set standards for the environmental conditions, floors and traffic routes, falls and falling objects, transparent or translucent doors, gates, walls and windows, safe cleaning of windows, sanitary and washing facilities, accommodation of breaks and eating food and changing work clothes.

In summary, design risk assessment can be illustrated as a design process using five steps to the risk assessment process where steps 2 and 5 represent the hazards and the design action respectively. This model also represents the steps or columns of a traditional risk assessment pro forma designed to record, coordinate and communicate health and safety issues for further action and auditing.

Providing information in or with design

The designer must think about the information that will assist the contractor or other designers in considering how hazards are to be managed at the design stage and not as a post-design health and safety exercise. Most of this information can be provided for referencing in the design risk assessment. An appreciation of the issues regarding access, site setup and establishment, and the construction and maintenance processes and interfaces will be invaluable in considering what information will be useful and in what format it should be provided.

CDM regulation 13(2)(b) states that every designer shall:

> ensure that the design includes adequate information about any aspect of the project or structure or materials (including articles or substances)

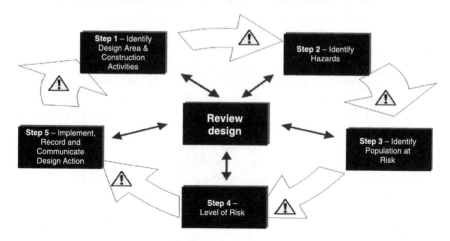

Figure 3.8 Five steps to risk assessment.

CDM Design Risk Assessment

Project-

Activity	Hazards	Population @ Risk	Risk			Design Action Taken, Mitigation & Risk Control Measures	Management of Action	
			L	S	RF		Further Action By (Name)	Associated Data Information

    Step 1   Step 2   Step 3   Step 4   Step 5

Figure 3.9 A typical risk assessment form.

which might affect the health or safety of any person at work carrying out construction work or cleaning work in or on the structure at any time or of any person who may be affected by the work of such a person at work.

The extent of the output of the risk assessment exercise is fortunately qualified by the guidance contained in the CDM Approved Code of Practice, which offers:

> Designers do not need to mention every hazard or assumption, as this can obscure the significant issues, but they do need to point out significant hazards.
>
> These are not necessarily those that result in the greatest risks, but those that are:
>
> • not likely to be obvious to a competent contractor or other designers,
> • unusual, or
> • likely to be difficult to manage effectively.
>
> CDM Approved Code of Practice

Competent designers can use this guidance to consider the information they are going to provide with their designs and what information will not add value and potentially obscure a significant piece of information.

Practically informing a roofing contractor there is a risk of falling off a roof is not going to help. However, information that will help should be based on the unknowns like the environmental issues and interfaces, the structural hazards and hazards associated with any bespoke materials or designs that the contractor may not have worked on before.

In Table 3.2 are some examples of unusual or difficult-to-manage hazards that support the provision of information.

Table 3.2 Examples of unusual or difficult-to-manage hazards that support the provision of information

Example	Type of information
A site with poor or no access from the road	Where possible, a temporary access route illustrated on the drawing.
Bespoke structural design to address an irregular shape	Provide information to support installation, e.g. pre-assembly instructions, installation methodology based on known structural environmental conditions for temporary works, weights of components to facilitate lifting plan development by contractor.
Installation of large beam into a confined/restricted space	Provide weight of beam and assumed access route or method for contractor so lifting plan, risk assessment and method statement can be developed.*
Interface with the public at a busy shopping centre	Define the extent of space allowed to work and lay down materials and equipment. Where possible consider reviewing the operating hours of certain construction or maintenance activities. For example, suggest that certain access equipment is not used during shopping hours.
Installation of partition walls where asbestos is present	Provide asbestos survey or commission/advise client to obtain information. Either recommend enabling works to remove asbestos or suggest alternative fixing method that avoids drilling/disturbing asbestos.
Installation of large curtain walling units over highway	Supplier/designer information on installation/lifting arrangements to reduce time by lifting section pre-assembled on site. Also designer could supply traffic and road closure information to facilitate development of risk assessment, and method statements regarding working over the highway.

(*Continued*)

Access for maintenance in a confined space/void	Highlight hazard to client for inclusion as a residual risk in health and safely file.
Component replacement – air handling unit on a building roof	Provide as installed drawings and lifting arrangements for inclusion in health and safety file. Where significant restrictions exist, provide outline method statement.
Specification of a hazardous substance as a finish	Provide hazard data sheet on the material and where applicable safe application instructions.**
Large eccentric truss	Provide centre of gravity guide for contractors/ appointed person*** planning a lift.
Demolition of buttress walls on an old part-occupied building	Provide existing surveys and drawings for temporary works designer and coordinator**** to consider when designing a solution.

* Residual risks can be transferred to the contractor where no reasonable design solution exists but to demonstrate the work can be undertaken safely supporting information to assist the operative in managing the risks is necessary.
** Those who specify products, materials and construction processes can act as designers. This also includes contractors selecting materials.
*** An appointed person is a competent person under the Lifting Operations and Lifting Equipment Regulations 1998 to manage a lifting operation.
**** A temporary works designer and coordinator are appointed to implement BS 5975: 1996, 'Code of practice for falsework'.

Examples of information that could be deemed obvious or unhelpful to a competent contractor are:

- informing a roofing contractor there is a risk of falling and falling objects;
- providing an electrician with a reference to the wiring regulations;
- informing a principal contractor that all workers are to wear hard hats;
- providing masonry contractors with a hazard data sheet on cement;
- stating on a design risk assessment pro forma that all site vehicles must have flashing beacons;
- requesting method statements for all activities on a design risk assessment;
- stating there is a risk in manually handling standard-size lintels;
- stating there is a risk of slips and trips and that the contractor must maintain high standards of housekeeping;
- informing a block layer that there is a risk of work-related upper limb disorder;
- providing a set of generic design risk assessments;
- informing a structural engineer that the design must avoid deep excavations;
- informing a steel frame contractor that there is a risk of structural collapse during the erection of a standard portal framed building.

The ACoP provides examples of significant hazards where designers always need to provide information. They include:

(a) hazards that could cause multiple fatalities to the public, such as tunnelling, or the use of a crane close to a busy public place, major road or railway;

(b) temporary works, required to ensure stability during the construction, alteration or demolition of the whole or any part of the structure, e.g. bracing during construction of steel or concrete frame buildings;

(c) hazardous or flammable substances specified in the design, e.g. epoxy grouts, fungicidal paints, or those containing isocyanates;

(d) features of the design and sequences of assembly or disassembly that are crucial to safe working;

(e) specific problems and possible solutions, for example arrangements to enable the removal of a large item of plant from the basement of a building;

(f) structures that create particular access problems, such as domed glass structures;

(g) heavy or awkward prefabricated elements likely to create risks in handling; and

(h) areas needing access where normal methods of tying scaffolds may not be feasible, such as facades that have no opening windows and cannot be drilled.

In summary, designers must provide adequate health and safety information in or with their designs on any aspect of the design that may affect the health and safety of any person at work or affected by the work. The extent to which this duty must be complied with is qualified by guidance in the Approved Code of Practice which only requests designers to mention unusual and difficult-to-manage issues and not those hazards that one would expect a competent contractor to manage effectively.

Cooperation with the planning supervisor and other designers

Designing on most projects requires a team to provide the necessary expertise in architecture, structural design, engineering, etc. Because the design elements interrelate, the team must coordinate the related elements of design such as cost, building regulations, aesthetics and also health and safety. The designer shall:

co-operate with the planning supervisor and with any other designer who is preparing any design in connection with the same project or structure so far as is necessary to enable each of them to comply with the requirements and prohibitions placed on him in relation to the project by or under the relevant statutory provisions.

CDM regulation 13(2)(c)

Cooperation with the planning supervisor and other designers so they can comply with their respective CDM duties obviously requires a sound understanding of their requirements and the mechanism used to coordinate and communicate health and safety information. The objective of cooperation is not only to avoid incompatibilities in design but to ensure hazards are managed effectively by the most suitably competent designer who is in control of that element of design contractually.

The main classification of information planning supervisors need is on residual design and project risks for collation in the pre-tender or pre-start health and safety plan. Without this information the planning supervisor cannot develop a suitable safety plan and cannot provide the client with information on the suitability of the construction phase plan for approval before work starts. To conclude, this information is essential if the planning supervisor is to comply with their respective CDM duties but also if value is to be added through the planning supervisory service.

In cooperating with other designers on the project information on residual risks and the environment, existing drawings and design in progress are essential. Information can be referenced on design risk assessments or on drawings.

Example: The structural engineer designing a mezzanine floor must have information on the existing or proposed building to design safely; likewise the mechanical engineer must know the layout of the structural elements so the pipework and ducting can integrate without affecting the structural integrity adversely.

The CDM Approved Code of Practice provides examples on how to demonstrate cooperation and how it can be encouraged by, for example:

- appointing a lead designer, where several designers are involved;
- agreeing a common approach to risk reduction during design;
- regular meetings of all the design team, contractors and others;
- regular reviews of developing designs;
- joint meetings to review designs, where there is a shared interest in an issue; and
- site visits.

In summary, designers must cooperate with the planning supervisor so they can produce a suitable safety plan and ensure designers are designing with adequate regard to health and safety. The provision of information to facilitate collaborative designing and design coordination meetings are recommended. The mechanism and tools necessary to promote and project-manage design coordination must be effective and based on the complexity of the design and project. There is a knock-on effect of not complying with this duty, and many designers in reality comply with this duty by default, as it is no more than good design practice which was in place long before CDM was introduced.

Good practice

To achieve the objective of safer designs there is an option to develop minimum design standards that have been based on a generic risk assessment. Policies like not specifying concrete blocks that weigh over 20 kilograms and low-maintenance products are typical examples of incorporating good health and safety into the design service.

The Health and Safety Executive promote a red, amber and green list system of in-house or project design practice. The lists are defined as follows.

Red lists

These are lists of known high-risk or hazardous building materials, construction activities and procedures to be eliminated from the project without compromise by designers. This list can also include CDM arrangements or CDM gateways for a project where no action can be taken unless a necessary statutory action has been executed.

Examples include:

- pre-tender health and safety plan not to be issued until detailed structural surveys, asbestos surveys, etc. completed;
- scabbling of concrete ('stop ends', etc.);
- demolition by hand-held breakers of the top sections of concrete piles (pile cropping techniques are available);
- the specification of fragile roof lights and roofing assemblies;
- processes giving rise to large quantities of dust (dry cutting, blasting, etc.);
- on-site spraying of harmful particulates;
- the specification of structural steelwork which is not purposely designed to accommodate safety nets;
- designing roof-mounted services requiring access (for maintenance, etc.), without provision for safe access (e.g. barriers).

Source: www.hse.gov.uk

Amber lists

These are products (as on the red list but lower-risk), processes and procedures to be eliminated or reduced as far as possible and only specified or allowed if unavoidable. Amber list items are by inclusion unusual and/or difficult to manage for other designers, the contractor or the facilities maintenance contractor and therefore require information to be provided. The list does not include the classification of hazards that one would expect a competent contractor or designer to be aware of, e.g. the risk of falling off a standard roof to a roofing contractor.

Examples include:

- internal manholes in circulation areas;
- external manholes in heavily used vehicle access zones;
- the specification of 'lip' details (i.e. trip hazards) at the tops of pre-cast concrete staircases;
- the specification of shallow steps (i.e. risers) in external paved areas;
- the specification of heavy building blocks, i.e. those weighing over 20 kilograms;
- large and heavy glass panels;
- the chasing out of concrete, brick or blockwork walls or floors for the installation of services;
- the specification of heavy lintels (the use of slim metal or concrete lintels being preferred);
- the specification of solvent-based paints and thinners, or isocyanates, particularly for use in confined areas;
- specification of curtain wall or panel systems without provision for the tying of scaffolds;
- specification of blockwork walls over 3.5 metres high and retarded mortar mixes.

Source: www.hse.gov.uk

Green lists

These lists promote a positive design-in culture and include products, processes and procedures that are to be positively encouraged on the project or as part of the in-house design standards.

Some of the advantages of this proactive and organised approach include an efficient management and decision-making process during design, which can otherwise cause designers concern when investigating solutions. Other advantages include a uniformity of approach. The lists and the arrangements for their management need to be flexible and evolve throughout projects to provide a continuously improving set of design standards and reduced risk for builders and maintenance operatives on the associated projects.

Examples include:

- adequate access for construction vehicles to minimise reversing requirements (one-way systems and turning radii);
- provision of adequate access and headroom for maintenance in plant rooms, and adequate provision for replacing heavy components;
- thoughtful location of mechanical and electrical equipment, light fittings, security devices, etc. to facilitate access and away from crowded areas;
- the specification of concrete products with pre-cast fixings to avoid drilling;
- specifying half-board sizes for plasterboard sheets to make handling easier;
- early installation of permanent means of access, and prefabricated staircases with handrails;

- the provision of edge protection at permanent works where there is a foreseeable risk of falls after handover;
- practical and safe methods of window cleaning (e.g. from the inside);
- appointment of a Temporary Work Coordinator (BS 5975);
- off-site timber treatment if PPA- and CCA-based preservatives are used (boron or copper salts can be used for cut ends on site).

Source: www.hse.gov.uk

Lead designer and lead consultant

There is no CDM regulation that requires a lead designer or consultant to manage and coordinate health and safety throughout the design process. However, the CDM Approved Code of Practice states that clients must, except for the simplest cases, set out in writing the arrangements for health and safety management. This includes designers and contractors correctly identifying hazards and control measures in accordance with regulation 13 of CDM and regulation 3 of the Management of Health and Safety at Work Regulations 1999.

These arrangements can form part of a formal contract or tender and, given that it is the designer's duty to ensure the client is aware of their CDM duties before starting design work, the designer would be advised to promote this formal approach to provide clarity on who is responsible for what.

An example of a client's or lead designer's safe designing or CDM project strategy is as follows:

- Decide on a project strategy to assist in the identification and design management of hazards.
- Consult with the client, design team and planning supervisor to form an agreement on the strategy.
- At the concept or feasibility stage, obtain all the client's information about the structure, environment and project.
- With the design team, undertake a brainstorming session or workshop to identify possible hazards and design interfaces. Record the findings and share information.
- Considering the client's information and initial brainstorming information, integrate a risk assessment process that further considers the issues in relation to the design development and in relation to significant hazards. Where possible, design to avoid, reduce, control or transfer the hazard. This is obviously related to the design and will continue throughout the project.
- Record only the actions required by others or information required, not the mitigation or assumptions if they are obvious on the drawing or in the specification.
- Hold stage design health and safety workshops or, for efficiency, integrate the subject of hazard management into the design team meetings for action

tracking or management and invite the planning supervisor so they can undertake their legal auditing function of the designer's health and safety inputs.

The Approved Code of Practice (ACoP)

To assist designers the ACoP provides guidance on the application of design risk assessment and explains the approach to designing with respect to the regulations as:

> The first stage in reducing risk is *to identify the hazards in a proposed design*. The next stage is to *eliminate each hazard, if feasible*. It is always best to design hazards out, so that no one is put at risk, but it is counter-productive to design out one hazard, only to introduce others that would result in a higher level of risk.
>
> Where it is not feasible to eliminate a hazard the next stage is to *consider what can be done to reduce the risk during the construction work* – this includes cleaning, maintenance or demolition. *In most cases it is sufficient to approach this using experience and published guidance, without sophisticated risk analysis techniques*. Designs which reduce the risks to everyone exposed should be used before turning to measures that only protect individuals – for example it is better to provide edge protection than rely on fall arrest systems.
>
> Finally, designers *must provide the information* necessary to identify and manage the remaining risks.
>
> CDM Approved Code of Practice

The ACoP also provides general examples of applied good design practice to assist designers and provide an opportunity to demonstrate, by selection, compliance with CDM regulation 13(2)(a). This list could form part of a strategy to design safely that one could argue is not a pure risk assessment but an alternative method of designing safely. The alternative approach to risk assessment will be discussed in Chapter 4. The examples include:

(a) Select the position and design of structures to minimise risks from site hazards, including:

- buried services, including gas pipelines
- overhead cables
- traffic movements to, from and around the site
- contaminated ground, for example minimising disturbance by using shallow excavations, and driven, rather than bored, piles.

(b) Design out health hazards, for example:

- specify less hazardous materials (e.g. solvent-free or low-solvent adhesives and water based paints)
- avoid processes that create hazardous fumes, vapours, dust, noise or vibration, including disturbance of existing asbestos, cutting chases in brickwork and concrete, breaking down cast in-situ piles to level, scabbling concrete, hand digging tunnels, flame cutting or sanding areas coated with lead paint or cadmium
- specify materials that are easy to handle (e.g. lighter weight building blocks)
- design block paved areas to enable mechanical handling and laying of blocks.

(c) Design out safety hazards, for example:

- the need for work at height, particularly where it would involve work from ladders, or where safe means of access and a safe place of work is not provided
- fragile roofing materials
- deep or long excavations in public areas or on highways
- materials that could create a significant fire risk during construction.

(d) Consider prefabrication to minimise hazardous work or to allow it to be carried out in more controlled conditions off-site including for example:

- design elements, such as structural steel work and process plant, so that subassemblies can be erected at ground level and then safely lifted into place
- arrange for cutting to size to be done off-site, under controlled conditions, to reduce the amount of dust release.

(e) Design in features that reduce the risk of falling/injury where it is not possible to avoid work at height, for example:

- early installation of permanent access, such as stairs, to reduce the use of ladders
- edge protection or other features that increase the safety of access and construction.

(f) Design to simplify safe construction, for example:

- provide lifting points and mark the weight, and centre of gravity of heavy or awkward items requiring slinging both on drawings and on the items themselves
- make allowance for temporary works required during construction
- design joints in vertical structural steel members so that bolting up can easily be done by someone standing on a permanent floor, and

by use of seating angles to provide support while the bolts are put in place

- design connections to minimise the risk of incorrect assembly.

(g) Design to simplify future maintenance and cleaning work, for example:

- make provision for safe permanent access
- specify windows that can be cleaned from the inside
- design plant rooms to allow safe access to plant and for its removal and replacement
- design safe access for roof-mounted plant, and roof maintenance
- make provision for safe temporary access to allow for painting and maintenance of facades etc. This might involve allowing for access by mobile elevating work platforms or for erection of scaffolding.

(h) Identify demolition hazards for inclusion in the health and safety file, for example:

- sources of substantial stored energy, including pre- or post-tensioned members
- unusual stability concepts
- alterations that have changed the structure.

(i) Understand how the structure can be constructed, cleaned and maintained safely:

- taking full account of the risks that can arise during the proposed construction processes, giving particular attention to new or unfamiliar processes, and to those that may place large numbers of people at risk
- considering the stability of partially erected structures and, where necessary, providing information to show how temporary stability could be achieved during construction
- considering the effect of proposed work on the integrity of existing structures, particularly during refurbishment
- ensuring that the overall design takes full account of any temporary works, for example falsework, which may be needed, no matter who is to develop those works
- ensuring that there are suitable arrangements (for example access and hard standing) for cranes, and other heavy equipment, if required.

CDM Approved Code of Practice

In summary, the CDM Regulations require designers to design with adequate regard to health and safety. The regulations appreciate that the designers can only achieve this to the extent that it is reasonable and practicable to do so, which introduces an element of judgement which is difficult to implement on a day-to-day design basis. The safe design duty also requires the designer to identify hazards and take action based on a hierarchy that promotes action to avoid hazards and where this is not practicable to design to reduce or control the hazards. The designer, in designing safely, also needs a strategy to apply the principles of risk assessment that should be based on the project's profile and promote the identification of hazards by identifying the known, and sometimes assumed, construction and maintenance activities. This strategy can also be considered as a design health and safety review agenda that should provide the designer with opportunities to address problems at the appropriate stage of the design.

The strategy and risk assessment mechanism must be proportionate to the nature and complexity of the design. Project arrangements set by the client, the designer or lead consultant must define the approach for the project if it is to achieve the legal and added value objectives. We must never forget that legislation, management tools and training on the CDM duties and associated hazards cannot by themselves create safe designs. Key to the success of designing safely is the designer's competence in providing a design solution to a health and safety problem, which requires the most valuable element of competence, namely experience.

Hazard and information management

The client is charged with the responsibility for the development of project CDM arrangements to define the roles and responsibilities and approach for compliance. The ACoP offers:

> Clients set the tone for the project and have overall control. Their approach is critical to how a project runs in practice. They do not have to manage projects themselves, but must ensure that there are clear and appropriate arrangements for managing and co-ordinating the work of all those involved. . . .

CDM Approved Code of Practice

Hazard management from the designers is a method of controlling the outputs and actions from a risk assessment exercise. Actions from the designer's perspective can include further design considerations or reviews, requests for information, meetings, product or materials research, etc. Their outputs also provide valuable information for construction safety and maintenance safety. So the output may sit in the construction phase health and safety plan or the health and safety file.

Depending on the arrangements, the lead designer or consultant is generally best placed to manage these hazards and their respective actions, but competent planning supervisors can, depending on their service level agreement, provide this additional valuable service. Tools like spreadsheets, risk registers and action-tracking software can assist in this process. In most simple cases the minutes of meetings can be adequate.

Developing competence as a designer

The competence referred to here is that which will satisfy the requirements of CDM regulation 8(2), which requires that no person shall appoint a designer unless he is reasonably satisfied that the designer is competent to undertake the design without contravening any statutory provisions, e.g. the CDM Regulations.

All professional designers, whether they be architects, architectural technologists, or civil, structural or services engineers, have to maintain standards of competence as part of their professional membership and status. The mechanism for this, in most cases, is continuous professional development (CPD).

Industry bodies, practice managers and educational establishments should by now be developing agendas for training that would benefit from covering the practical elements of designing safely. Workshop training on live projects delivered by experienced designers and planning supervisors would provide many organisations with a platform for learning on their own projects.

Training criteria for any programme should focus on the mechanics of integrating design risk assessment in its simplest and purest form into the design process and service. The training should cover examples of good and bad design to illustrate the value of CDM as no more than good design and not a bolt-on issue involving the completion of paperwork. Finally, the training should cover the interactions with the other CDM duty holders and how the hazards are managed throughout their project or the structure's life through the primary CDM documents.

Some frequently asked designer questions

Q. *Do I have to record all my design actions regarding health and safety?*

A. No. The ACoP explains that this is not a legal requirement, but on certain

X Project Design Risk Register						
Hazard Ref No	Hazard Ownership	Design Area/ Construction/ Maintenance Activity	Hazard Issue	Design Action to Date	Further Design Action	PS Comments & Supporting Information
AR1.1	A Company	Ground / Foundations	New structure clashing with existing piles	Carry out accurate survey of existing pile locations to determine relationship with new structure.	Survey existing piles. Fix setting out of new building in world co-ordinates. Co-ordinate existing piles with new structure.	??????? to arrange structural design CDM review meeting before end of May. (Deadline for PT H&S Plan)

Figure 3.10 Project risk register.

complex design elements within projects it may be advantageous to communicate to other designers the reasoning behind a design solution where there is a design interface. All you have to record is information on hazards with significant risks that are unusual or difficult to manage, not hazards and assumptions that a competent designer or contractor could easily manage. Many of the design actions are evident on drawings, in specifications and in the minutes of meetings.

Q. *What reasonable steps should I take to inform a client of their duties under CDM when they will not discuss these issues with me or respond to my letters?*

A. There are no hard-and-fast rules regarding informing clients of their duties. However, following the ACoP is advisable. It advises the designer to draw the client's attention to the first three sections of the ACoP and the HSC leaflet called 'Having construction work done?' The document also advises designers who are dealing with inexperienced clients to emphasise the importance of the client's role, good health and safety management and making early appointments. This can be done during early meetings or by providing briefings to the client.

In providing a demonstrable contribution, a standard letter and minutes of meetings can be valuable. Please note that designers are only required to take reasonable steps to advise clients before they start design work, and intervention is not required after that point. However, again it is added value to advise the client as the project develops.

Where clients are obviously ignoring the advice and breaching the regulations the designer should not feel compromised but be confident that they acted professionally in providing advice as required.

If a client was investigated by the enforcing authority the provision of additional CDM tools to assist the client in executing their duties would clearly demonstrate the designer's commitment to taking reasonable steps to advise the client.

From a designer's or practice's perspective, compliance with CDM regulation 13(1) should be a default procedure that is automatically addressed on every project.

Q. *How do I generally demonstrate that I design with adequate regard to health and safety?*

A. Many designers are concerned about proving their contributions by recording every action. This is not reasonable, and the ACoP provides two clear examples to give designers confidence in this respect. Firstly, the ACoP explains that designers are not legally required to keep records of the process they use, commonly known as design risk assessments, and secondly the ACoP explains that designers don't have to mention every hazard or assumption as this can obscure the significant issues.

Demonstrating that you have designed safely is a combination of design inputs and outputs. Inputs like spatial design, specifying lighter-weight materials and avoiding an interface with an existing unstable structure can be clearly seen and demonstrated on drawings and in specifications. Outputs in the form of information on hazards and supporting information for future management of the hazards during construction and maintenance can also be recorded on the drawing or, where more information and future action need managing, a risk assessment or risk register template can be used. This output information could include construction or maintenance methodologies for unusual or difficult-to-manage operations, and lists of hazards that one would not expect a competent contractor to be aware of.

The recording of certain design actions can be advantageous when the complexity of the design, the interfaces and the research mean that this information may be required by other designers, e.g. on design-and-build contracts.

Q. What does CDM regulation 13(2)(a) (designing with adequate regard to health and safety) actually want me to do?

A. The requirement to design safely is derived from the Management of Health and Safety at Work Regulations 1999 and appreciates that not all hazards can be eliminated or reduced and that in relation to some hazards there is a cost to designing safely. These regulations also have a requirement to apply the principles of prevention and protection in a hierarchical order.

Terminology like 'adequate regard' and 'reasonably practicable' requires the designer to make a balanced and holistic judgement based on the probable outcome of a hazard causing an accident or ill health against cost, programming, buildability, function, sustainability and aesthetics. The regulations and enforcers are not looking for a total hazard elimination process. However, designing with adequate regard to health and safety could be interpreted as requiring the designer to eliminate foreseeable risks where possible and to reduce the level of risk for the residual risks and make them easier to manage for the contractor and maintenance operatives. The method of achieving this is hazard management, and an integral element of this process is risk assessment.

Q. What health and safety-related information do I and don't I include with my designs?

A. The information should be based on risk and complexity. If an element of the design is complex then the provision of information in or supporting the design will help communicate the design parameters to the other designers, suppliers and contractors. Conversely a simplistic design does not warrant information for simple structural or construction elements and will only contribute negatively to the service.

The ACoP also lists the qualifications for providing information:

Designers do not need to mention every hazard or assumption, as this can obscure the significant issues, but they do need to point out significant hazards.

These are not necessarily those that result in the greatest risks, but those that are:

- not likely to be obvious to a competent contractor or other designers,
- unusual, or
- likely to be difficult to manage effectively.

CDM Approved Code of Practice

The ACoP also lists the types of significant hazards where information should always be included. They are:

(a) hazards that could cause multiple fatalities to the public, such as tunnelling, or the use of a crane close to a busy public place, major road or railway;

(b) temporary works, required to ensure stability during the construction, alteration or demolition of the whole or any part of the structure, e.g. bracing during construction of steel or concrete frame buildings;

(c) hazardous or flammable substances specified in the design, e.g. epoxy grouts, fungicidal paints, or those containing isocyanates;

(d) features of the design and sequences of assembly or disassembly that are crucial to safe working;

(e) specific problems and possible solutions, for example arrangements to enable the removal of a large item of plant from the basement of a building;

(f) structures that create particular access problems, such as domed glass structures;

(g) heavy or awkward prefabricated elements likely to create risks in handling; and

(h) areas needing access where normal methods of tying scaffolds may not be feasible, such as facades that have no opening windows and cannot be drilled.

CDM Approved Code of Practice

Q. *How far do I go with risk assessment and providing information? Is there a rule of thumb?*

A. This depends on the nature of the project and the complexity of the design. As one rule of thumb, designers should:

- identify all significant hazards by analysing the possible construction and maintenance methods and where possible design to avoid them; or

- where this is not possible try to reduce the risk of the remaining hazards to make them either negligible or easier to manage on site. This may include providing information in or with the design on the residual hazards.

The in-house arrangements and system used to apply this strategy should also dictate how far one goes with risk assessment. The system can be a simple hazard checklist or where necessary a risk register.

As another rule of thumb, designers can also use the regulations themselves, which provide and qualify how far one can be expected to go to comply with the duty to design safely:

> the design to include only the matters referred to therein to the extent that it is reasonable to expect the designer to address them at the time the design is prepared and to the extent that it is otherwise reasonably practicable to do so.
>
> Extract from CDM Regulation 13(3)

In terms of carrying out a legal duty, the term 'reasonably practicable' means that the degree of risk in a particular activity can be balanced against the time, trouble, cost and physical difficulties of taking measures to avoid the risk.

Q. *I have completed a competence and resources questionnaire, did not get the project, but would like some feedback, which is not forthcoming from the planning supervisor. What can I do?*

A. The planning supervisor does not have a legal duty to disclose their report on your questionnaire, competence and resources. However, in the interests of continuous improvement and professionalism the planning supervisor or the design practice could seek the permission of the client to disclose the report.

One of the problems in assessing the competence and resources of designers is the naturally subjective nature of quantifying and qualifying the information and how to avoid the planning supervisor making a personal judgement without sufficient information. Professionally organised planning supervisors should have a scoring system and in the interests of the industry should publish results and where necessary recommendations.

Q. *How should I cooperate with the planning supervisors and other designers?*

A. One of the obvious answers is to communicate with them. The project arrangements for design management and coordination should provide a suitable platform to address health and safety issues. On some projects planning supervisors provide a design health and safety coordination service, which is not a legal requirement, but when the lead designer is the planning supervisor this has obvious advantages of efficiency and the demonstration of an integrated approach.

The extent to which a designer must cooperate with the planning supervisor and other designers is such that they can comply with their duties. The main designer's duties concern designing safely, so information on your design that is unusual and difficult to manage would be helpful. Planning supervisors' duties mainly concern the auditing of the designer's health and safety contributions, and the provision of design risk assessments, hazard identification checklists, etc. would in most circumstances provide them with sufficient evidence.

On most projects, having the planning supervisor attend design team and review meetings could be considered a waste of resources. However, if the planning supervisor is not provided with evidence in the form of risk assessments then there is sometimes little alternative. Lead designers could be proactive and invite the planning supervisor to design reviews where an agenda item could be the risk register or indeed CDM.

Q. *What CDM role do lead designers play?*

A. There is no specific, statutory CDM role for a lead designer, but the client is responsible for setting the arrangements for the management of health and safety for the project and the lead designer is contractually responsible for coordination of all aspects of the design so why not health and safety? Lead designers are in the best position to coordinate the health and safety aspects of the design and integrate them into their service.

Based on their design competence, management position in the project and resources, there is a strong argument for them to deliver the planning supervisory service, a notion that the HSE's proposed revisions to the CDM Regulations have potentially picked up on.

Q. *How do I know if I am complying with my duties?*

A. A competent designer would by definition know if they are complying with their designer duties. To measure your compliance you need to assess how you have addressed your four primary duties.

In respect of advising the client of their duties, have you taken reasonable steps and provided your client with information in the form of the HSE leaflet 'Having construction work done?' and made them aware of the first three sections of the ACoP? For less experienced clients, have you emphasised the role of the client, good health and safety management and the advantages of making early appointments of the key CDM duty holders?

In respect of designing with adequate regard to health and safety, have you identified all the significant hazards associated with the construction and maintenance and designed to eliminate these where possible? Where this is not possible, have you designed to reduce the risk to make it more manageable for the contractor to work safely? And have you informed the contractor of significant residual risks that a competent contractor would find difficult to manage effectively?

In respect of providing information in or with your designs, have you provided information where the ACoP states through examples that information must always be provided in respect of significant hazards? Also does the information address the requirement not to mention every hazard or assumption as it can obscure the significant issues? Designers do need to provide information on unusual and difficult-to-manage hazards but not on hazards that one would expect a competent contractor to manage effectively.

And finally, in respect of cooperating, have you provided the planning supervisor and other designers with project health and safety information so they can comply with their respective duties?

If the answer to these four issues is yes, then you are complying. If the answer is no, then you need to revise your project CDM contributions.

Q. *How do I know if I am a competent designer as required by the CDM Regulations?*

A. The definition of competence in relation to a designer in the regulations only covers the requirement to comply with relevant statutory provisions and undertake any requirements. The regulation actually states that:

> No person shall arrange for a designer to prepare a design unless he is reasonably satisfied that the designer has the competence to prepare that design.
>
> CDM regulation 8(2)

The following is a list of issues one could use as a general checklist to define designer competence:

- recognised qualifications in the particular area of design;
- membership of a professional body;
- experience in the particular area of design;
- a working knowledge of the CDM Regulations;
- a clear understanding of hazard management processes.

In summary, if you or your practice have addressed the above issues then you are competent under CDM.

Q. *I have engaged another design company to work with me on a project. What do I have to do under CDM?*

A. Any person, including another designer, who engages a designer must be reasonably satisfied that they are competent and adequately resourced. So you can either:

- ask the planning supervisor to offer advice on the competence of the designer before they start; or
- undertake your own assessment which could include:

  interview

  issuing a standard questionnaire

  obtaining information on previous assessments during the project (for novation and design-and-build contracts).

Note that a design-and-build contractor who takes on the existing design team assumes responsibility for their respective competence and provision for health and safety.

The extent to which one should go to be reasonably satisfied that they are competent should be based on the risk profile of the project. On a high-risk, complex design project a more competent designer will be required to ensure the hazards are managed effectively. On a simple building extension, simple refurbishment project or small domestic house the designer will only need to be sufficiently competent to address the health and safety implications likely to be involved with the design.

Q. A client is overruling my decision to install a parapet in favour of a man-safe system. What do I need to consider?

A. Legally the client is acting as a designer and is responsible for that design decision. As the client's CDM adviser, legally required to advise the client of their duties before starting design work, you would be well advised to extend this service to advising them of their legal status in making design decisions, even though it is not a duty. Designers are placed in many compromising positions on projects where their designs are externally influenced. It is important to know that a designer is not always the architect or engineer but anyone who makes a design contribution, and that includes the client. The only deviation from this regulation applies to persons acting in a legal capacity, e.g. a planning officer giving a planning direction or a building control officer ordering compliance with a specific building regulation. This is not to say that these directions do not affect health and safety on a project, but it is the designer's responsibility to consider the health and safety impact of the issue.

Sometimes design standards are imposed on designers, e.g. a housing association specifying a type of window that cannot be cleaned from the inside and must be located in a difficult-to-access location on a house. In this case the person or organisation that specified the window could be seen as the responsible designer.

In all these cases, record the communication and professionally advise the client on the grounds of best practice. Where a lead designer uses a hazard management matrix and allocates design management responsibility for elements of the design, the client can be issued with responsibility to clarify their duties under CDM as a designer.

Q. *I am starting a complex project and the design-and-build contractor, as part of my contract, wants our design studio to act as lead designer with respect to CDM. I have to give a presentation on our proposed arrangements to the client. Where do I start?*

A. The client is responsible for setting up the arrangements for CDM but, in arranging for the design-and-build contractor to contractually require the architect to act as lead designer, one could argue that the client is acting properly.

The main issues that design practices need to emphasise in the presentation on acting as lead designer from a CDM perspective are:

- Standard setting. This could be achieved by having a basic training Power-Point presentation or workshop presented by an experienced designer on the arrangements of hazard management including the CDM tools and reporting requirements. Also the most common reference and guidance books and material should be brought to the attention of the design team, for example *Safety in Design (SID)* produced by the Construction Industry Council, the CDM Approved Code of Practice, and *Work Sector Guidance for Designers* by CIRIA. The standards set should also include roles and responsibilities for all the design team with the specialists leading in their particular discipline.
- Hazard management. This is about the arrangements and tools that will be used to design safely, including arrangements for design coordination. It should address the type of risk assessment pro forma or risk register and the rules for its use, including what is not to be recorded and the rules for the provision of additional information in or with the design, e.g. notes on drawings, supporting design information where necessary to assist in the on-site management of health and safety, e.g. the weight or lifting arrangements for a heavy and awkward material, the installation arrangements for an unusual or new product that is difficult to install in the given environment, and advice on the liaison and management arrangements when the site is occupied during construction, e.g. the phasing of works, etc.
- Design review and audit. Here one could define the arrangements in respect of the timing and design stages, the criteria and content of the audit, and the formal procedure to analyse the health and safety contributions to design out problems and design in solutions. This is generally achieved using an audit checklist.

The key to the presentation is getting the design team's commitment and also the commitment of the client to support the statutory and added-value potential of CDM. Other issues that could add value to the presentation are demonstrating that CDM is an integrated element of the design service and demonstrating with examples how you add value. Examples could include M&E systems' interface with the structure, off-site construction or low-level assembly, easy-to-access and -maintain elements of the structure or building showing lower maintenance costs.

Q. *We are about to be investigated by the Health and Safety Executive following an accident on a building site where we were the engineering designers. What should I do?*

A. The Health and Safety Executive on a routine inspection of a design practice look for three main elements that they need to see to give them the confidence that the practice or individual designer has the potential to comply with CDM.

The three main areas on which the HSE concentrate during inspections are:

- CDM policy and procedures;
- training, covering:

  legal background
  health and safety background (knowledge of site risks)
  hazard management (knowledge of identifying and designing to avoid hazards and designing to reduce residual risk)
  resources (HSE books, publications of design safety practice, etc.);

- application of CDM (examples on previous projects where hazards have been designed out or reduced and the provision of suitable information to facilitate health and safety management by the principal contractor).

In respect of a specific accident the above information will act as valuable mitigation during the investigation. The HSE will look for evidence that the hazard or hazards associated with the construction activities were considered and what action was taken or could have been taken at the design stage.

Q. *The local planning officer has commented on my design and is preventing the contractor incorporating a one-way traffic management system into the site setup. What action should I take?*

A. The officer does not have a duty under the CDM Regulations because they are acting in their capacity as an authorised officer under the respective legislation. However, if there is no obvious argument for not implementing the designer's recommendations and there is a significant increase in risk in not having a one-way system it may be worth arguing in writing on the grounds of health and safety or public safety. Town and country planning officers, highways planning officers and other enforcement officers have to act within their powers and have a duty of care under civil legislation, which sometimes needs highlighting to promote the safest way forward.

Q. *The project client is not allowing me sufficient time to consider and manage health and safety as a part of the design service. What should I do?*

A. The pace and poor planning on some projects is an issue that obviously affects all design service elements. In most cases where time is scarce it will not be possible to consider the health and safety aspects in detail. However, there are still contributions that can be made that would only take a competent professional designer a few minutes, e.g. drafting a list of unusual or difficult-to-manage hazards.

However, the main issue here is a failure of the client or project manager to properly plan and manage the project. The CDM ACoP advises that the client should have arrangements in place to allow adequate time for design, planning, preparation and construction work.

Q. *I cannot think of any serious health and safety problems that a competent contractor could not manage effectively, but the planning supervisor and client are insisting on design risk assessments showing all thinking behind my design decisions. What should I do?*

A. The issue is that you have no output from the health and safety management element of your design. There are no unusual hazards or difficult-to-manage hazards. You also are not required to provide information on hazards that one would expect a competent contractor or designer to be able to manage effectively. So you are not legally required to address their request.

However, this request may be based on compliance with a set of CDM project management arrangements, or the information may be required because the planning supervisor is not aware that it is not a legal duty to produce risk assessments but the planning supervisor misguidedly uses them as evidence of designer compliance.

If you are required contractually on the project to provide a design risk assessment, a list of hazards considered should be sufficient. If they are not significant they will require no further action. Where some issues or design options were considered it may be worth recording them, but not completing the pro forma with generic statements.

A concern that CDM auditors are faced with is that organisations and project teams go through the process of asking for CDM documents and actions and this, in the opinion of some, demonstrates that CDM is being managed. But the objectives and practical reasoning behind CDM are hazard management, and just because a risk assessment pro forma has been issued to a planning supervisor does not mean that the designer has designed to their optimum potential.

Q. *A planning supervisor has told me that they would expect at least 15 pages of design risk assessments for a project of this size. Is this right?*

A. The quantity of health and safety output information should be based on unusual and awkward hazards that a competent contractor may need to know, and does not require the designer to mention every hazard or assumption, as it can obscure the significant issues.

Another factor that can define the output is the rule that the designer is required, through project arrangements, to show their full risk assessments and mitigation to the planning supervisor so they can assess the designer's contributions to ensure they are designing with adequate regard to health and safety as required by the CDM Regulations.

The main factor that should determine the designer's health and safety input and outputs is the design complexity. On many low-risk, simple design projects there are no residual design risks that one would not expect a competent contractor to manage effectively. But on complex projects with many design contributions the coordination and collaboration of design health and safety matters will be extensive and can contain many significant residual risks that the contractor will need to plan and manage proactively.

Chapter 4

The planning supervisor

Introduction

The original European Communities Directive 92/57/EEC on which the United Kingdom legislators based the CDM Regulations called for the client to appoint a 'coordinator for health and safety matters'. This appointee's key objective would be to coordinate health and safety matters during the various stages of designing and preparation for a project. Consequently the UK lawmakers invented the 'planning supervisor' to fulfil this role.

The introduction of the planning supervisor was seen, amongst the more cultured and forward thinking in the construction industry, as a huge opportunity to improve pre-construction health and safety. This was achieved through early involvement with the designers, ensuring that hazards were being considered and eliminated and subsequent risks reduced. Sceptics, however, believed that the creation of the planning supervisor introduced an unnecessary tier of bureaucracy and even a duplication of duties that were already undertaken by the lead designer, lead consultant, project manager or contract administrator.

The name 'planning supervisor' has never sat well with many in the industry, as they argue that this duty holder neither 'plans' nor 'supervises'. It does, therefore, appear that the name of this particular duty holder will disappear when the CDM Regulations are next reviewed as the role evolves. Much time and effort have gone into debating this relatively trivial issue, whilst many in the industry still do not actually appreciate the project value of a competent planning supervisor in terms of risk elimination and reduction. The fact, however, remains that for construction projects where CDM is applicable the client must appoint a planning supervisor.

Who then is best placed to assume the role of the planning supervisor and audit the health and safety issues likely to be involved on a given project? The client needs to ensure that the planning supervisor is competent and adequately resourced for the project. A selection of skills and knowledge required should include:

- knowledge of health and safety in construction, including all relevant legislation;
- a thorough understanding of the respective design processes;
- knowledge of the various construction processes and materials used in construction;
- an understanding of what designer and contractor competence to design and construct with respect to the statutory provisions is and how to evaluate it;
- a sound understanding of how construction health and safety management should be implemented;
- good interpersonal and communication skills.

Planning supervision is undertaken by lead designers, architects, engineers, health and safety professionals, project managers, clients and even the principal contractor (on design-and-build schemes, for example).

The planning supervisor can be an individual or an organisation. The planning supervisor can also be an existing member of the project team or completely independent. Many believe that best practice would see the planning supervisor as an independent appointment in that they would be more willing, arguably, to challenge the design team's health and safety contributions. The regulations, however, simply demand that the appointed planning supervisor is competent and adequately resourced, but how a client would do this is not clear, given that many clients are lay people.

For a planning supervisor to be fully effective their appointment must be during the earliest design stages or even pre-design if they are to be in a position to advise on the competence and adequacy of resources of the design team. The late appointment of the planning supervisor has been a major failing in the industry and regulations to date. All too often a great deal of design work is undertaken without the skills of someone to holistically assess the health and safety implications of a project's construction, future maintenance and even eventual demolition or dismantling when most of the opportunities exist to make fundamentally safer design decisions.

Another essential requirement for any planning supervisor that adds value is to ensure the design team cooperate and work as a team in their health and safety management function. The planning supervisor will need to ensure that the design team, for example, are comfortable with the arrangements for health and safety management at the respective design stages which ideally will be integrated into the entire design management process. The skills of communication and ability to present the case for an inclusive role will generally lead to a harmonious relationship with the design team. The planning supervisor does not wish the project team to view them simply as a CDM policeman and must strive for cooperation.

Not surprisingly, an entire planning supervision industry has sprung up on the back of the CDM Regulations. Several planning supervision 'clubs' were established to clarify the role, offer guidance and training, and establish fee levels for aspiring 'coordinators for health and safety management'. It is important for clients to know that membership of any such organisations is not necessarily a guarantee of competence.

Before we look in detail at the actual functions of the planning supervisor the point must be made that, in terms of these duties, they hold significant *potential* to contribute to a well-designed and planned scheme from a health and safety management perspective. But you will note that in real terms the planning supervisor has very little statutory influence and generally is required to delicately articulate the benefits of their inclusion in the major decisions taken.

Significant lost opportunities can occur in terms of successful health and safety planning when planning supervisors:

1 are appointed late;
2 are not competent;
3 are not adequately resourced;
4 do not posses adequate resources;
5 do not receive client support;
6 are not provided with relevant information from the client;
7 are not provided with relevant information from designers;
8 fail to secure a working and integrated relationship with designers.

It is fair to say that the above list of planning supervisor issues is all too common as the construction industry still tries to get to grips with this key health and safety management opportunity. Many claim that the role of the planning supervisor has been mis-sold to the client, who often sees the position as an unnecessary overhead and additional red tape.

However, although clarity as to the role of the planning supervisor clearly exists in such publications as the Approved Code of Practice and CIRIA Report 172, planning supervisors are still widely perceived as CDM facilitators and *not* as 'coordinators for health and safety matters' as defined in the original EU directive.

> The planning supervisor's main responsibility is to ensure that those who carry out design work on a project, particularly during the design phase, collaborate and pay adequate attention to the need to reduce risk wherever possible.
>
> CDM Approved Code of Practice

Many planning supervisors provide the client with a service level agreement which provides the client with the option to select the service or not. In a competitive construction environment where planning supervisors are selected on their fees or the service is thrown in as part of the project management, a minimalist approach is sometimes delivered. A minimalist service to a client can ensure the client and planning supervisor are addressing their CDM duties but adds no value in real terms. A minimalist service can involve no direct contact with the designers and client and is managed administratively with questionnaires, standard letters and pro formas.

The planning supervisor's duties explained

We will next explore the respective duties of the planning supervisor in detail and demonstrate, in a more positive light, how they can influence and contribute significantly to health and safety on site, and beyond.

Notification to the Health and Safety Executive

If the client, designer or planning supervisor, depending on project arrangements, has established that the project meets the criteria for notification, in that the construction phase will be longer than 30 working days or that the construction phase will involve more than 500 person days of construction work, they must ensure that the Health and Safety Executive are informed. Note that the planning supervisor must only 'ensure' that this is done and does not necessarily have to undertake the task personally. This is a typical duty that, unless the client's CDM arrangements are in writing, could lead to confusion. In most cases the planning supervisor undertakes this task. The standard HSE form for notification is called the F10 and is available on the HSE's website where it can be sent electronically.

The details that need to be provided to the Health and Safety Executive are defined in Schedule 1 of the regulations themselves:

Schedule 1: Particulars to be notified to the Executive

1. Date of forwarding.
2. Exact address of the construction site.
3. Name and address of the client or clients.
4. Type of project.
5. Name and address of the planning supervisor.
6. A declaration signed by or on behalf of the planning supervisor that he has been appointed as such.
7. Name and address of the principal contractor.
8. A declaration signed by or on behalf of the principal contractor that he has been appointed as such.
9. Date planned for start of the construction phase.
10. Planned duration of the construction phase.
11. Estimated maximum number of people at work on the construction site.
12. Planned number of contractors on the construction site.
13. Name and address of any contractor or contractors already chosen.

Notification needs to be made as soon as is practicable following the appointment of the planning supervisor. If not all of the details for the initial notification are known then a second notice must be sent to the Health and Safety Executive. It is often the case on a traditional procurement route that the principal contractor will not be known until post-tendering. Therefore a second or additional notification is often required following the appointment of the principal contractor.

The planning supervisor shall ensure that notice of the project in respect of which he is appointed is given to the Executive in accordance with paragraphs (2) to (4) unless the planning supervisor has reasonable grounds for believing that the project is not notifiable.

CDM regulation 7(1)

Finally the planning supervisor is the gateway for ultimately deciding whether the project falls within the scope for notification and this decision can be difficult when the qualifying criteria are borderline, e.g. number of person working days and total number of working days. Many projects planned for 30 days can overrun and, at the time of deciding, the brief or extent of the design may not be finalized. Experienced planning supervisors err on the side of caution and notify, which is good practice.

Advising the client on the health and safety competence and resources of the team

The planning supervisor has another duty to be available, if required, to offer advice to the client on the competence and resources of any proposed designer or contractor they wish to employ for a project. The planning supervisor also needs to be in a position to offer advice to any contractor regarding the competence and resources of any designer they may wish to engage.

The planning supervisor appointed for any project shall –

. . .

(c) be in a position to give adequate advice to –

  (i) any client and any contractor with a view to enabling each of them to comply with regulations 8(2) and 9(2), and to

  (ii) any client with a view to enabling him to comply with regulations 8(3), 9(3) and 10.

Extract from CDM regulation 14

The problem with this duty is that most designers are actually employed or under instruction to design before the planning supervisor is appointed so the assessment of their competence is retrospective and not an integrated part of the procurement route.

If the client is not especially competent to assess the competence and resources of the team and has not appointed a client's agent, then it is prudent to seek the advice of the planning supervisor. The client, remember, must be 'reasonably satisfied' in terms of assessing competence and resources, so the opportunity to embrace the planning supervisor's advice and experience can facilitate compliance with these duties and ensure the designers and contractors are competent to fulfil their legal obligations under CDM.

This is one regulation where the planning supervisor can be all too ineffective if the client does not appreciate the valuable contribution they can potentially make during the appointment of the team. Their input at this stage can assist the client in 'setting the tone' for health and safety management.

Case study: A client wished to invite three demolition contractors to tender as principal contractor, and the planning supervisor was asked to offer advice on competence and resources. It was evident to the planning supervisor that one of the contractors was not competent to address the health and safety issues likely to be involved in the scheme. The planning supervisor mentioned his concerns formally to the client, in writing. However, the client was adamant that the contractor be invited to tender and ignored the advice of the planning supervisor.

On the return of the completed tenders the contractor in question came in with a considerably lower price than the two competent demolition contractors. It was also evident to the planning supervisor that the contractor was not adequately resourced to manage the health and safety issues. The other two contractors clearly satisfied the requirements in terms of demonstrating health and safety resources for the demolition work. The planning supervisor again made the client aware of his concerns. The client thanked the planning supervisor for his comments but went on to engage the cheapest contractor and proceeded to offer them a contract to act as principal contractor for demolition work. The contractor was served with two improvement notices within the first month of work and the client received numerous letters from the public and environmental health department concerning dust and noise nuisance.

The fundamental strategies for assessing competence and resources are based on a competent person who understands the project issues and can assess and measure an organisation's and individual's competence and resources against those project-specific issues. It could be argued that without a benchmark or target to make the assessment the exercise is too subjective and a waste of time. The issuing of generic pre-qualification questionnaires (PQQs) is suitable for developing tender lists, but unless specific questions on issues associated with the type of project are introduced you will never be in a position to assess competence to comply with the CDM duties on a particular project, which is the objective of the exercise. For in-house planning supervisors, the arrangements for the assessment should be seamlessly integrated within the procurement procedures, which avoids increased bureaucracy.

The interview and the questionnaire are the two methods planning supervisors have at their disposal to undertake the task. On a simple, low-risk project, selecting a designer or contractor who has previously completed the questionnaire will generally be seen as sufficient evidence for the planning supervisor to advise the client.

On more complex projects an interview or project-specific questionnaire is advisable. The interview should involve the interviewee answering questions regarding specific difficult-to-manage issues on the project for an assessment of competence, e.g. 'What design considerations would you consider regarding the interface with the party wall and maintaining access through the site?', 'In relation to the installation of the steelwork what design considerations would you make to reduce the risks associated with installation and structural stability during installation?' The practical problem with this exercise in some cases is that the designer within a design company has not been identified and only the organisation's general competence can be assessed. In this case it is valuable to put a comment in the report to the client that the lead designer and design team require further assessment.

Note: Planning supervisors in executing their duty to advise the client on competence and resources should in all but the simplest of cases measure competence and resources against the health and safety demands of the project, which should be assessed by them initially.

In relation to assessing resources, obviously human resources are a big part of the assessment, but technical resources must also be covered. If an organisation is under-resourced it will potentially not be in a position to address its statutory obligations. Designers will not be able to attend all meetings and workshops and may not have time to consider health and safety. Contractors will potentially not have the safety professionals to manage site safety or implement the conditions of their construction phase health and safety plan. As with competence, this assessment is based on the planning supervisor's ability to judge what is needed to deliver the design service or construction service to the standards of the CDM Regulations and Approved Code of Practice.

Finally the reporting of the planning supervisor's findings should be presented in an advisory style and in some cases will identify that a designer or contractor is or is not competent or adequately resourced to address their CDM responsibilities on a project. However, the planning supervisor may be generally confident following their assessment but may think there are some weak areas in, for example, evidence of training or previous experience, etc. In this case it is good practice, as part of the advisory service, to recommend action that the design practice or contractors could take to strengthen their application. As a rule of thumb, always qualify the assessment report and state why a designer, design practice, principal contractor or contractor is or is not competent and adequately resourced. This professional response to the CDM Regulations adds value to the clients in respect of their business or project risk assessment.

All the above factors are ultimately governed by client decisions, and the resource of time, one of the most valuable resources in managing health and safety, is often a precious commodity. A designer on competitively low fees and a compressed design programme cannot be expected to perform as diligently from a CDM perspective as a designer with plenty of lead-in time and adequate fees for the project. To this end some procurement routes that follow the traditional competitive tendering processes must set minimum standards and a level playing field for competitors if they want to be rewarded with the CDM and other construction benefits of good practice.

Ensuring that adequate regard has been given to health and safety during design

This is arguably where the planning supervisor can play the most significant role in terms of their contribution to reducing a project's risk profile. This is also where the client will find out whether the assessment of competence and resources was effective and whether the client's interpretation of the advice was heeded. The ability to work closely with designers from an early project stage to ensure that hazards are identified and eliminated or reduced can be a vital contribution to on-site health and safety. Risks associated with future maintenance and even eventual demolition or dismantling of the structure can also be designed out or significantly reduced following involvement from the planning supervisor.

> Planning supervisors have to be satisfied that designs address the need to eliminate and control risks as required by regulation 13(2).
>
> CDM Approved Code of Practice

Early on, the planning supervisor will need to consider an appropriate strategy to carry out this duty effectively. This is generally dictated by the size of the project in terms of the number and variety of designers involved, the procurement route and the planning supervisor's own perception of the level of risk involved.

The main routes to assessing the designer's health and safety contributions are:

- Attending design review meetings and assessing the designer's contributions and health and safety considerations. In these design environments where the design options are discussed with the client and other consultants the planning supervisor can, but is not legally required to, enter the discussions on design solutions that avoid or reduce the hazard and add value.
- Requesting design risk assessments from the design team unless already issued to the planning supervisor in compliance with their duty to cooperate. Evaluation of these pro formas should provide sufficient evidence of the designer's health and safety contributions or failings. These documents should be read and assessed with the drawings and or specifications. This also depends on whether the arrangements for recording outputs of design risk assessments include mitigation or just future actions and the provision of information.

- Calling a CDM design review meeting to formally discuss the designer's contributions and strategies. This provides the ultimate opportunity to explore all aspects and interfaces of the design and the respective designers' health and safety considerations. This environment provides the opportunity for the planning supervisor, lead designer or lead consultant responsible to manage health and safety, through a risk register for example.

  Planning supervisors may need to encourage or arrange design review meetings if they are not satisfied there is sufficient interaction between designers or if adequate regard is not being given to health and safety. Such meetings can identify unforeseen problems.

  CDM Approved Code of Practice

- Reviewing design meeting minutes and contributions.

In assessing the designer's contributions, the planning supervisor is also assessing their competence in the field of design health and safety and based on this initial assessment the planning supervisor can evaluate their strategy to further assess the designer's contributions. For example, if a designer has not identified deep excavations in a restricted area as a significant risk will they identify the access issues in relation to the plant located on the roof?

  Planning supervisors who identify important health and safety issues that have not been addressed in a design should draw them to the attention of the designer.

  CDM Approved Code of Practice

As with all duties 'to ensure', the planning supervisor, in the report to the client, project manager or client representative, should highlight areas of concern that have not been addressed fully by the design. These residual risks or issues concerning the planning supervisor should be of a nature where it is reasonably practicable for the designer to have taken design action to avoid or reduce the risk to a more manageable level on site or during maintenance.

The extent of this duty is qualified by 'so far as is reasonably practicable'. This does not require the planning supervisor to evaluate all aspects of the designer's health and safety contributions but should provide a basis for a strategy that is based on the project's risk profile. It would be adequate on a smaller, lower-risk project to adopt a minimalist strategy, whereas on a large, complex design and high-risk project a more detailed and investigative approach which explores the design areas and construction activities in more detail would be advisable.

Ensuring that adequate information is provided in or with the design

This designer's duty should be considered by the planning supervisor as an integrated element of the requirement to design with adequate regard to health and safety and is merely an output of a design risk assessment. If a hazard needs consideration by another designer or a method statement requires action to be taken by a maintenance contractor, the designer should clearly communicate this information with the design.

Again, for the planning supervisor to be in a position to ensure the designer is providing adequate information in or with the designs an assessment of the project and the presumed information should be made first. Designers do not need to provide information on every hazard or assumption but on the unusual and difficult-to-manage issues. The planning supervisor shall:

> (a) ensure, so far as is reasonably practicable, that the design of any structure comprised in the project –
>
>   . . .
>
> (ii) includes adequate information as specified in regulation 13(2)(b).
>
> Extract from CDM regulation 14

Note that the design risk assessment is valuable supporting information that should be provided with the design. Typical information a competent contractor should be providing with their designs includes detailed drawings for difficult-to-assemble or -replace elements of the building or structure, and information on residual maintenance hazards, e.g. confined spaces, heavy components, difficult-to-access light fittings, etc. Many of these hazards should be supported by a method statement on how the hazard can be managed safely.

It is invaluable for a planning supervisor, who is one of the legal parties required to receive this information, to have it provided in a manageable and collated format. The planning supervisor should use this information, provided throughout the project, to complete the pre-tender health and safety plan and monitor the future actions of designers within the team as part of the planning supervisor's auditing process. This information can also provide valuable evidence of designers cooperating in respect of health and safety if challenged.

Ensuring cooperation between designers

On all but the simplest of projects, design development is a team effort and, as with all aspects of design, health and safety requires a collaborative approach. Initially to comply with this duty, the planning supervisor can set up arrangements to monitor cooperation with the client and, depending on the design's complexity, the approach can vary from reading design review meeting minutes to attending design team meetings. But monitoring is only the means to an end, and the regulation states that the planning supervisor shall:

> (b) take such steps as it is reasonable for a person in his position to take to ensure co-operation between designers so far as is necessary to enable each designer to comply with the requirements placed on him by regulation 13;

CDM regulation 14(b)

As stated in the regulation, this duty extends to taking such steps as are reasonable for a person in the planning supervisor's position to ensure cooperation between designers. Depending on the planning supervisor's position, e.g. they could be lead designer, consultant or an independent planning supervisor with limited resources, the approach obviously requires a judgement and would benefit from clarification in a service level agreement with the client. Tools and methods to facilitate cooperation can vary from informing the lead designer or consultant, with examples, of one's concerns, to design programme Gantt charts highlighting hazards, resources and design responsibility, and risk registers leading with hazards and recording design action taken and future design action by whom and when.

Planning supervisors with limited design or service experience need to be aware that competent, experienced designers naturally cooperate with other designers in solving all types of design problems and issues. It should not be assumed that, because there is not a formal design lead strategy for cooperation, the designers are not complying with this duty. Design experience and experience in contributing to design team meetings where issues are debated and resolved are invaluable for any planning supervisor in discharging their duties efficiently and adding value.

In assessing the designer's health and safety contributions and CDM compliance the following is a list of the issues the planning supervisor should be looking for in holistically evaluating a designer's health and safety input:

- Clear evidence of the project designers identifying significant hazards at different stages of the design, e.g. an architect considering site access and special design/footprint layout options at concept and feasibility stages and M&E interface with the structure during the scheme and detail stages of design.
- Evidence that the designers have a strategy, whether formal or not, for considering hazards associated with the construction processes and assumed construction processes. The strategy should demonstrate to the planning supervisor that the designers will consider hazards at all stages of

the design and potentially review the design health and safety contributions at the recognised service stages.

- Evidence that design actions have been taken to avoid, reduce, control and transfer hazards.
- Evidence that designers are cooperating in respect of their design contributions to avoid, reduce, control and transfer hazards. This cooperation may involve simple arrangements to define the zones for the service runs within a false ceiling void. The structural engineer and services engineer will need to work together to avoid design conflicts. Other problems could involve access during construction with badly planned service runs and potential access issues for maintaining plant.
- Evidence that information and actions, as outputs of their risk assessment, are being managed by the respective recipients; and information for additional design contributions, collated in the health and safety plan by the planning supervisor and included on the health and safety file for future maintenance considerations. Evidence that designers know how the residual hazard is to be managed and by which CDM duty holder is valuable and in most cases essential.

Ensuring that a pre-tender health and safety plan (the plan) is prepared in good time

The information provided by the client on the project, structure and environment coupled with the residual construction hazards derived from the designer's risk assessment forms the basis of health and safety information to be provided to a principal contractor post-contract. In providing a contractor with information on the difficult-to-manage site issues, environment and client's requirements, the contractor has time to plan and resource for health and safety before starting work on site. Historically this information if available is held within the tender documentation, but CDM requires this data to be formalised in a CDM document.

> The planning supervisor appointed for any project shall ensure that a health and safety plan in respect of the project has been prepared . . . .
> Extract from CDM regulation 15(1)

This document, prior to the construction phase of a specific project, is generally referred to as the 'pre-tender' or 'pre-start' health and safety plan.

Any planning supervisor who fully understands the legal significance of the Approved Code of Practice will generally follow its recommended layout for the pre-tender plan. The level of detail provided in the plan should also be relevant to the risk profile of the project. The plan's contents should contain information on the following where applicable:

1. Description of project

    (a) project description and programme details
    (b) details of client, designers, planning supervisor and other consultants
    (c) extent and location of existing records and plans

2. Client's considerations and other management requirements

    (a) structure and organisation
    (b) safety goals for the project and arrangements for monitoring and review
    (c) permits and authorisation requirements
    (d) emergency procedures
    (e) site rules and other restrictions on contractors, suppliers and others e.g. access arrangements to those parts of the site which continue to be used by the client
    (f) activities on or adjacent to the site during the works
    (g) arrangements for liaison between parties
    (h) security arrangements

3. Environmental restrictions and existing on-site risks

    (a) safety hazards, including

        boundaries and access, including temporary access
        adjacent land uses
        existing storage of hazardous materials
        location of existing services – water, electricity, gas, etc.
        ground conditions
        existing structures – stability, or fragile materials

    (b) health hazards, including

        asbestos, including results of surveys
        existing storage of hazardous materials
        contaminated land including results of surveys
        existing structures – hazardous materials
        health risks arising from client's activities

4. Significant design and construction hazards

    (a) design assumptions and control measures
    (b) arrangements for co-ordination of on-going design work and handling design changes
    (c) information of significant risks identified during design (health and safety risks)
    (d) materials requiring particular precautions

5.   The health and safety file

   (a) format and content.

CDM Approved Code of Practice, Appendix 3

Much of the detail to be provided for the pre-tender plan formally comes from the client and the designers as part of their duty to provide relevant health and safety information to the planning supervisor. This is often critical information that contractors require if they are to be able to consider and manage health and safety risks that are over and above those which can be expected for the type of project and respond to the plan effectively by producing a construction phase health and safety plan. If either the client or a designer does not provide information, the consequences are often absorbed by the contractors and can result in site problems.

Case study: A designer specified 30-metre trusses for a new supermarket to be constructed. The site had only a single-lane access. The client had not informed the planning supervisor of the restricted-access issue and consequently the designer was also unaware. This detail was consequently omitted from the pre-start health and safety plan.

When the trusses arrived they could not access the site and remained parked on a main road causing a major traffic jam. The police came and instructed the trusses be moved on. The contractor had to erect a much larger tower crane and lift the trusses into place from the main road as an overnight operation. This proved to be a much more hazardous (and costly) operation.

The golden rules of producing a suitable and sufficient pre-tender health and safety plan are:

- Don't include information which one would expect a competent contractor to be able to manage effectively.
- Don't include general generic statements about health and safety standards.
- Include design and site information that will facilitate the contractor in undertaking construction risk assessment and developing method statements to promote safe systems of work on site.
- Set standards for what is expected by the client and planning supervisor in respect of health and safety management. A principal contractor and contractor will benefit from clarity as to what is acceptable in relation to the development of a construction phase health and safety plan.

Planning supervisors historically have added value to projects by liaising with the client in respect of their site activities where applicable, undertaking investigations with the architect and carrying out site visits to obtain more information.

The planning supervisor needs to examine all the available information to identify the key health and safety issues for the plan. In particular, issues that have significant resource implications need to be clearly identified. Any client requirements which have significant implications for health and safety must be clearly stated.

Planning supervisors may request information from others to include in the plan, but usually collate, manage and issue it themselves. The plan should be issued with the tender documents, so that those tendering have all necessary information.

Approved Code of Practice Guidance

The development of the pre-tender health and safety plan provides an opportunity to review the client's information and design information and, as part of the duty to ensure the plan's development, to request additional information. The types of information one would expect a client to provide could include existing drawings of services or newly commissioned services surveys, examples of the client's permits to work for occupied sites, asbestos registers and even a current health and safety file for the structure.

If, in the increasingly bureaucratic construction world, planning supervisors expect principal contractors to respond appropriately to the pre-tender health and safety plan, a minimalist and succinct approach with clarity on what is expected to be approved by the client is essential. As the client or planning supervisor must assess the construction phase health and safety plan and deem it suitably developed for work to start, providing a contractor with the benchmark in the form of a concise pre-tender plan with clear project issues that need responding to either within the plan or by supporting method statements and risk assessments will pay dividends.

Advising the client on the initial contents of the construction phase health and safety plan

This duty again requires the planning supervisor to provide advice and assist the client with their holistic health and safety management responsibilities. To this end a planning supervisor must have the skills to assess a safety document that complies with the Approved Code of Practice, addresses the issues raised in the pre-tender health and safety plan and demonstrates the potential and resources to manage health and safety and coordinate contractors' health and safety on site. As with the duty to advise on competence and resources, the regulations only require this service to provided if requested.

Every client shall ensure, so far as is reasonably practicable, that the construction phase of any project does not start unless a health and safety plan complying with regulation 15(4) has been prepared in respect of that project.

CDM regulation 10

Numerous clients have been successfully prosecuted for allowing construction work to begin without a suitably developed construction phase health and safety plan (or indeed without a construction phase plan at all!). This is an opportunity for a client to take guidance on a significant CDM duty.

If a planning supervisor is keen to add value and further contribute to the health and safety management process, this service should be offered on all but the simplest projects. It could be possible that the client may have sufficient competence to ensure that the construction phase health and safety plan demonstrates compliance with regulation 15(4). If this is the case then the planning supervisor's experience may not be called upon. However, there are few examples of such clients in reality, and planning supervisors should employ a proactive approach to making themselves available to offer advice.

It is also true to say that planning supervisors can put themselves in a relatively compromising position if they suggest to the client that the construction phase health and safety plan is unsuitable for construction work to start. Traditionally, the client is enthusiastic to begin on site and will not welcome any potential delays if the principal contractor needs to review the standards set in the plan. So, early intervention by the planning supervisor working with the client or project manager in setting lead-in times and a project programme for the CDM milestones can prove valuable.

Furthermore, the planning supervisor can play an additional proactive role in working with the principal contractor, if necessary, to ensure that the construction phase health and safety plan is suitable and sufficient and that there are no delays or conflicts. This untraditional approach can be further expanded into developing what one could call an integrated health and safety plan that contains information from the pre-tender and construction phase health and safety plan in one document. This approach, if undertaken accurately, does not compromise compliance with the CDM Regulations and for many in-house organisations can prove far more efficient.

Case study: Within a couple of days of the expected start of work on site, the principal contractor emailed the construction phase health and safety plan to the planning supervisor for consideration at the request of the client. The plan was clearly not suitably developed and did not satisfy the regulations. It included no site-specific information such as an emergency plan or a site traffic management plan.

The planning supervisor had to inform the client, who was extremely anxious that the project should start on site, that the plan did not, in his opinion, meet the requirements of regulation 15(4) and consequently work should not start. Although the client was far from happy with the news, the planning supervisor managed to articulate the statutory responsibility and the plan was returned to the principal contractor for improvement.

In reporting findings to the client, the planning supervisor would benefit from using a pro forma to demonstrate that all the plan's contents have been assessed and also to set comments against any weak areas and recommend further development.

The comments on the plan can be weighted to emphasise to the contractor the importance of the omission. For example, if the name of a consultant has not been included in the plan, a simple note and no further action are satisfactory. However, if there are no arrangements in the plan for temporary works design and coordination on a large highway concrete retaining wall then the comment would advise the client not to allow work to start until the issue is ratified within the plan and reassessed by the planning supervisor.

In summary, there are four general objectives the construction phase health and safety plan must achieve for the planning supervisor to deem it suitably developed for work to start. The plan must:

- address the significant residual hazards identified by the designer;
- address any issues identified by the client in respect of site rules, restrictions and interfaces;
- demonstrate the potential and resources for the principal contractor to manage health and safety and coordinate health and safety issues between contractors throughout the construction phase;
- have arrangements to collect health and safety file data and deliver it as specified to the planning supervisor for assessment and delivery to the client.

Note: The starting of construction work without a suitably developed construction phase health and safety plan both practically and legally is a high-business-risk issue for the client. CDM regulation 10 does not have an exemption from civil liability like other CDM regulations. It could be argued that, based on this, the advice given to the client in respect of the plan is the most valuable on a project.

Ensuring that a health and safety file is collated and handed over to the client

Health and safety people have argued for a long time about the risks during construction as against the risks during future maintenance. Statistically you are six times more likely to be killed on a construction site than in most other jobs but, as a product of the building's life, the maintenance operations and management of associated hazards go on for the structure's life. Therefore, the CDM document that provides the information to facilitate the future health and safety management is a valuable resource. All maintenance and future projects would benefit from a suitably developed and accessible health and safety file. Legally the client needs to ensure that the file is available for inspection. The regulations regarding the file initially state:

The planning supervisor appointed for any project shall –

...

(d) ensure that a health and safety file is prepared in respect of each structure comprised in the project containing –

   (i) information included with the design by virtue of regulation 13(2)(b), and

   (ii) any other information relating to the project which it is reasonably foreseeable will be necessary to ensure the health and safety of any person at work who is carrying out or will carry out construction work or cleaning work in or on the structure or of any person who may be affected by the work of such a person at work.

Extract from CDM regulation 14

The CDM Regulations initially require the planning supervisor to ensure that the health and safety file is prepared. The are no regulations that define who should collate the file so, as with many of the regulations, the client would be wise to include the arrangements for the provision, collation and approval of information and production of the file.

The planning supervisor is best placed to monitor the file's development through the audit function of monitoring the outputs of design risk assessments and information provided in or with designs. The planning supervisor also has the opportunity to define the arrangements for the file's development in the pre-tender health and safety plan. Certain complex projects require development of and access to the file throughout the project. Other projects require parts of the file to be handed over to tenants or occupiers when parts of a building are handed over for fit-out by other or the respective client contractors.

In setting up the file, the planning supervisor could consider doing the following, which could facilitate development:

- Set up a checklist of typical information for the designers and contractors, e.g. types of components that will be difficult to replace, such as air handling units, large window or glass sections, etc.; access arrangements, e.g. mobile elevation work platforms, access latchway systems, etc.; as installed M&E drawings, as built drawings, hazard data sheets for hazardous substances like finishes or materials used, contact details of specialist contractors, etc.

- Where possible, discuss the operational and maintenance health and safety issues with the client's health and safety representative, and review risk assessments and accident reports.

- Discuss the format of the file with the client and offer advice on version control and accessibility with information technology options.

Summary

In conclusion, the planning supervisor can, if competent and adequately resourced, add real value to the project and client's investment. If the client does not understand the function and value of this service then a minimalist approach may be adopted and the process of complying with the legislation may be achieved, but the likelihood of designer- and contractor-led health and safety problems is significantly increased.

It is fair to say that the industry has not understood the value of the planning supervisor, and even competent ones have struggled to change the culture on projects from the 'this is the way we have always done it' attitude. The planning supervisor was never really intended to be a stand-alone service but merely a set of functions that the existing project team could perhaps facilitate. To this end it would appear that the planning supervisor as a position is to be removed from the existing revision of the CDM Regulations and replaced with a set of coordinator functions to be addressed by whoever the client deems competent and suitably positioned, from within the project team perhaps. It is the author's opinion that, however professional a project team or organisation is, the commercial pressures on projects coupled with a lack of client sympathy and understanding have not provided a platform for competent planning supervisors to thrive. As with many aspects in design, unless a legal gateway is established designers as managers of technical standards on a project will find it difficult to apply principles of risk assessment and design health and safety standards that are based on a reasonable approach rather than a more prescriptive one.

Chapter 5

The principal contractor

Introduction

Up until the appointment of the principal contractor and start of work on site, all efforts in the proactive management and design of health and safety issues should have focused, with the exception of maintenance work, on the environment and the construction activities the contractors are charged with managing. In respect of this, for complex projects many procurement advisers to clients recommend a more integrated route to take advantage of the contractor's knowledge. These two-stage and design-and-build contracts engage the contractor at the design stage and if properly coordinated provide the potential for all in the project team to consider buildability, materials selection and the sequence and construction programme from a health and safety perspective with the benefit of experience from the construction site.

The principal contractor is ultimately responsible for the management and coordination of health and safety throughout the construction phase of a project. They are at the sharp end of the project and challenged with managing the residual design and environmental hazards. They are very much removed from the comfort of the design studio and the client's offices. The theory of a well-planned and -designed project would suggest that many significant risks will have been eliminated or reduced pre-construction, and the principal contractor will be required to manage the residual hazards on site. If the client, designers and planning supervisor have been diligent and carried out their own CDM responsibilities effectively, the rewards should be evident at this stage. If they have not, then unusual and difficult-to-manage health and safety issues on site or for future maintenance could be identified by a competent contractor, but the opportunity for a design change could be lost. The difference between a principal contractor having time to proactively manage a planned interface with the client's staff, e.g. the moving of offices, planned client maintenance work on site, etc., and finding out on the day can significantly affect work and reduce the opportunity to undertake a suitable and sufficient risk assessment to avoid or reduce risks as required.

A competent principal contractor will legally and contractually control their site. This control from a health and safety perspective and all the associated standards and arrangements should be defined specifically for the project in the construction phase health and safety plan.

It is valuable to appreciate that the term '*principal* contractor' is purely applicable to the statutory responsibilities of this duty holder as set out in the CDM Regulations. The principal contractor, therefore, is a legal entity in its own right and is not to be confused with the main contractor or the contractor.

It could be argued that the CDM Regulations have had less impact on organisations carrying out the role of principal contractor, as many contractors, particularly the larger companies, were already familiar with the principles of risk assessment, hazard management and health and safety legislation. CDM did, however, bring a framework to construction health and safety management that has been largely welcome by contractors and provided the more cultured of them with the opportunity to discuss health and safety with clients and designers and formally request information using the framework of responsibilities imposed on the other duty holders.

It is generally, although not always, the case that the main contractor carries out the duties of the principal contractor. They are obviously well placed to holistically coordinate and manage health and safety during construction. However, there are examples of clients themselves either acting as principal contractor or even arranging for a project management team to fulfil the role. Other examples of how the role is discharged by clients involve changing the principal contractor throughout the project so that the main or largest contractor takes responsibility based on the fact that they are most competent in managing the risks associated with their work. Best practice would dictate that for continuity and clarity one principal contractor throughout the project would have more opportunity to set, implement, monitor and review health and safety standards.

Because of the potential for appointing any organisation to act as principal contractor, e.g. management consultants or a health and safety company, a specific regulation is in place to ensure that only contractors act as principal contractors. This is the only CDM duty that must be undertaken by a specific organisation. As long as people or organisations are competent and adequately resourced, anyone can be a designer, client, planning supervisor or contractor under the CDM Regulations.

> The client shall not appoint as principal contractor any person who is not a contractor.
>
> CDM regulation 6(2)

Example: A hotel company that has a construction and maintenance division to maintain its hotels has experience of fitting out hotels and undertaking small construction projects. On new hotel projects it employs the main contractor building the structure as the principal contractor. The decision is based on the contractor's competence and resources to better execute the principal contractor's duties.

Any client engaging a principal contractor will benefit from considering the added value a competent and adequately resourced organisation brings to a project. If problems are encountered in obtaining site information as part of the client duties or if a designer omits to provide information on a difficult-to-access area of the site and if this type of information is omitted from a pre-tender health and safety plan then a competent contractor will still manage the issue. Competent contractors that are adequately resourced manage safety proactively with risk assessment exercised and plan the work by producing method statements with planned health and safety actions of identifying services, sample soil conditions, liaising with existing occupiers and third parties, etc. Under-resourced and less competent principal contractors do not manage safety proactively, and the effects of this impact on the client, designers and planning supervisors from a CDM compliance perspective.

Principal contractors fail to discharge their duties effectively for many reasons, not all of which are obviously under their control. Possible causes, which should be considered when setting up a contract, include:

1   Principal contractors are appointed so late that they cannot plan for safety.
2   They are not competent to manage health and safety and do not have suitably competent health and safety advisers, management or supervisors with sufficient knowledge, experience and training to develop project-specific arrangements.
3   They do not possess adequate resources to set, implement, monitor and audit the appropriate health and safety standards for the project.
4   They do not receive client support when construction programmes are extended.
5   They are not provided with relevant information from the client.
6   They are not provided with relevant information from designers on unusual or difficult-to-manage hazards or issues or are provided with irrelevant information on hazards that a competent contractor could manage effectively that could obscure significant issues.
7   They fail to secure a working relationship with designers.
8   They fail to secure a working relationship with and control subcontractors.
9   There is a lack of control on site, with clients controlling their own operations without the principal contractor's input or knowledge.

In summary, the principal contractor is a legal CDM position with the sole purpose of securing the health, safety and welfare of anyone who works on or resorts to the site or third parties who may be affected by the operations on or adjacent to the site. As for all CDM duty holders, there are specific CDM duties for the principal contractor and general CDM duties that a principal contractor must comply with, e.g. the appointment of competent and adequately resourced contractors and designers.

The principal contractor's duties explained

The duties of the principal contractor can be summarised as follows:

- selection of competent and adequately resourced contractors and designers;
- the construction phase health and safety plan;
- ensuring cooperation between contractors;
- controlling site access;
- displaying the project notification;

- enforcement of the site rules;
- provision of relevant information to contractors;
- health and safety training and information;
- the health and safety file.

Appointment of contractors and designers

As we have noted, if the principal contractor is to appoint contractors or designers they must ensure that the contractors and designers are both competent and adequately resourced to address the health and safety issues likely to be involved in the project. This duty requires a judgement to be made based on an assessment of information obtained from the contractor that can be collected in an interview, via a questionnaire, or both.

(2) No person shall arrange for a designer to prepare a design unless he is reasonably satisfied that the designer has the competence to prepare the design.

(3) No person shall arrange for a contractor to carry out or manage construction work unless he is reasonably satisfied that the contractor has the competence to carry out or, as the case may be, manage, that construction work.

Extract from CDM regulation 8

(2) No person shall arrange for a designer to prepare a design unless he is reasonably satisfied that the designer has allocated or, as appropriate, will allocate adequate resources to enable the designer to comply with regulation 13.

(3) No person shall arrange for a contractor to carry out or manage construction work unless he is reasonably satisfied that the contractor has allocated or, as appropriate, will allocate resources to enable the contractor to comply with the requirements and prohibitions imposed on him by or under the relevant statutory provisions.

Extract from CDM regulation 9

The principal contractor must avoid an overly bureaucratic approach to assessing competence and resources but at the same time needs to be confident that they have sufficient evidence of any designer's or contractor's ability to comply with their legal health and safety responsibilities. The amount of information requested or checks made should be proportionate to the risks likely to have to be managed. One of the key factors here, as with so many of the CDM regulations, is the competence of the person making the decisions. Assessing competence and resources is by its very nature subjective, so the strategy must have essential performance indicators to measure or benchmark organisations being assessed. In respect of competence the key issues are:

- Does the contractor have *adequate training and skills* to manage the health and safety aspects of the contract activities safely and do the work safely to the appropriate health and safety standards?
- Does the contractor have *evidence of experience* on this type of work?
- Has the contractor *failed to comply* with their duties or had formal action taken by the HSE?

In respect of resources:

- Does the contractor have *sufficient human resources* in the particular areas necessary to execute the contract safely?
- Does the contractor have *sufficient technical resources and equipment* at their disposal necessary to execute the contract safely?

Many designers have traditionally found it particularly difficult to demonstrate competence as there are limited or no accepted standards like the contractors' Construction Skills Certification Scheme (CSCS) or the Construction Industry Training Board's Site Management Safety Training Scheme. The revision of the CDM Regulations has potentially promoted a review of this situation, with the proposed new ACoP having guidance on competence and resources.

Principal contractors are not legally required to keep records and evidence of this process; however, it could be considered good practice if the process were subject to an investigation by the HSE.

A balanced and appropriate approach to this duty could include a questionnaire with options of more in-depth questions that could provide the assessor with the opportunity to vary the request for information from the designer or contractor. On a small, low-risk project only five questions with supporting information will apply, as opposed to 20 for a complex project. For a complex design project, a designer could be asked for their written procedures of how health and safety information is managed and coordinated between designers throughout the project. On a relatively low-risk project, a copy of their risk assessment pro forma could be deemed suitable to make the assessment.

Any CDM duty holder must be 'reasonably satisfied' that any appointee is both competent and adequately resourced to undertake their statutory health and safety obligations. To this end a balanced approach is essential, and many organisations have contributed to the industry charges made on the CDM Regulations that they are bureaucratic and do not contribute to health and safety as intended.

The assessment of competence and resources if undertaken professionally by a competent person can set the right tone for the project, even before it starts. One could argue that if this regulation and supporting industry information on compliance were more rigorously addressed and enforced it would only raise health and safety standards.

The construction phase health and safety plan

The vehicle used to manage and coordinate the health and safety aspects of the construction phase is the construction phase health and safety plan. It should be seen as the master health and safety policy for the project on site. As with a health and safety policy, the document should have the objectives of the plan, called the statement of intent in a statutory health and safety policy, the organisation for its implementation and the project-specific arrangements for managing health and safety on site. The construction phase health and safety plan, as with any health and safety policy, must also have clear arrangements for audit and monitoring procedures. The arrangements should include reactive aspects, i.e. accident and near-miss investigations, and proactive aspects, i.e. on-site monitoring and audits of management systems, safe systems of work and reacting to changes in design, etc.

The principal contractor must develop the health and safety plan throughout the construction phase of the project. The plan should build on the pre-tender or pre-start plan information, describe just how health and safety will be managed and coordinated during construction and clearly demonstrate how any project-specific health, safety or welfare issues will be addressed.

The level of detail in the construction phase health and safety plan will need to be proportionate to the project's specific risk profile. The ACoP suggests a list of the minimum information that should be in place from the start of the construction work, which can be used as a guide when developing plans and management systems and when discussing the development of a plan with a client or planning supervisor. The minimum information is listed as:

(a) general information about the project, including a brief description and details of the programme;
(b) specific procedures and arrangements for the early work;
(c) general procedures and arrangements which apply to the whole construction phase, including those for the management and monitoring of health and safety;
(d) welfare arrangements;
(e) emergency procedures; and
(f) arrangements for communication.

These initial requirements are further extended in Appendix 3 of the ACoP 'where they are relevant to the work proposed'. The headings below act as a checklist to assist the principal contractor when planning for health and safety. The headings below are supported by guidance on application that is not in the ACoP.

1. Description of work

a. Project description and programme details

Guidance on application

Simply include a brief description or overview of the project description covering the basic construction principles and based on the nature of the project programme details can vary from a detailed Gantt chart to start and finish dates for a simple project.

Objectives

To provide contractors and others who may interface with the project with an understanding of what the project is and its extent and sufficient information to consider the impact of the project on their activities e.g. client, maintenance team on site etc. or contribution to the project e.g. start of demolition, opening of new road, removal of asbestos etc.

b. Details of client, planning supervisor, designers, principal contractors and other contractors

Guidance on application

Full contact details of all parties including subcontractors can be helpful so all readers can contact any party as necessary. Arrangements with sub-contractors should include providing this information and any planned changes to the key personnel for safety which can also be checked at induction by the principal contractor.

Objectives

To provide up to date information and contact information on key health and safety personnel to promote and facilitate health and safety cooperation and coordination.

c. Extent and location of existing records and plans

Guidance on application

This information could include an existing health and safety file for an existing structure, reports of surveys undertaken and information on how to access this information if it is not contained within the plan. Most of this information should be referenced in the pre-tender health and safety plan which should have derived from the client making reasonable enquiries about the existing site or structure. Important information to have in or reference to in the plan is existing services surveys, structural or conditional surveys, asbestos surveys and existing drawings or method statements of historical construction works.

This information is invaluable when the principal contractor starts undertaking risk assessments for the project and setting up arrangements for induction training, briefing on the safety plan and providing information to subcontractors so they can plan for safety.

A competent principal contractor should also question any potential omissions and feed back to the client any concerns they may have regarding health and safety.

Objectives

To provide a focus for the principal contractor in setting site safety standards and information to support the selection of contractors and approval of method statements before work starts. Also information on significant unforeseen hazards e.g. buried services, structural defect etc., that contractors will need to consider in undertaking risk assessments of their activities and in developing method statements.

2. Communication and management of the work

a. Management structure and responsibilities

Guidance on application

The implementation of the construction phase health and safety plan requires the key health and safety skills and responsibilities clearly defined here. A list of appointments and/or an organogram or family tree best illustrates the hierarchy. The extent of detail on key personnel and their duties obviously depends on the size and nature of the project. On a small project, simply recording the construction managers', site supervisors' and/or senior tradesmen's names and responsibilities will be adequate. On large, more complex projects, with numerous design and contractor packages perhaps, the structure may need to be as illustrated below:

- construction manager/project manager;
- health and safety adviser;
- design manager/coordinator;
- purchaser;
- site supervisors;
- site foremen;
- appointed person (lifting operations);
- permit controller;
- temporary works coordinator;

- security manager;
- fire safety coordinator;
- first-aiders.

Duties and responsibilities can start with the initial and continuous development and overseeing of the implementation of the construction phase health and safety plan and can be as simple as maintaining the first aid provisions on site. The duties should address the arrangements for implementing the health and safety plan, emergencies and any contingencies.

Objectives

To clearly define the organisation to implement the construction phase health and safety plan and manage health and safety on site.

b. Project health and safety goals and arrangements for monitoring health and safety performance

Guidance on application

In respect of project health and safety goals it is valuable to define the objectives of the project in respect of health and safety. Examples of these could be:

- to ensure an accident and injury-free project;
- to have a zero tolerance on non-compliance with the site health and safety rules;
- to have weekly/monthly health and safety initiatives to promote a positive health and safety culture;
- to operate a 'Stop me/Don't walk by' policy to promote contractor participation in maintaining high health and safety standards;
- to operate safety incentive schemes, e.g. hazard busters, where site personnel are encouraged to report hazards for a prize or money;
- to contribute to national health and safety schemes, e.g. working well together;
- to set specific key performance indicators for high-level elements of the construction phase health and safety plan, e.g. method statement receipt and approval, permit management, access platform inspections, etc.

All safety targets must be realistic, achievable and promoted by the senior management.

For the monitoring of health and safety on site, arrangements need to cover:

- the person responsible for auditing and monitoring;

- the nature and frequency of audit;
- the management of output action from the audit.

And the fundamental targets for auditing and monitoring compliance should cover:

- the relevant health and safety regulations applicable to the project;
- the applied arrangements for selection and management of contractors;
- the selection and management of designers' health and safety contributions;
- the site rules contained within the construction phase health and safety plan;
- accident, dangerous occurrence and near-miss reports;
- hazards reported under the hazard-stopping incentive scheme;
- high-risk environments, e.g. working platforms, cofferdams, confined spaces, highways engineering, deep excavations, roof work, working over or near water, contaminated sites, etc.;
- high-risk activities, e.g. lifting operations and cranes, piling, live electrical work, asbestos removal, traffic management on site, using cartridge-operated power tools;
- training, results of awareness tests, induction tests, etc.;
- feedback from toolbox talks and talking to contractors at work during audits;
- method statements and risk assessments;
- permits to work;
- lifting plans;
- site coordination registers.

Monitoring of health and safety standards is generally undertaken by safety advisers and managers but involvement of all the team is good practice and promotes health and safety throughout the project team. The use of standard pro formas to prompt the auditor and record the findings is effective for site monitoring, and more management system-focused pro formas are used for auditing the implementation of the construction phase health and safety plan. Accidents and near-miss investigations are also invaluable tools to assess the effectiveness of the plan. Certain inspections of working platforms, cofferdams, excavations, etc. should be undertaken again with a checklist or pro forma but recorded on a formal register to assist in the management of the exercise. Systems like 'scaff tag' (a safety sign label that is attached to scaffolding to record the inspection and status of the scaffolding) are used to communicate the results of inspections and facilitate the management and outputs of the exercise. If scaffolding fails a weekly inspection the sign will read 'Do Not Use' with a red prohibition sign.

The extent to which the principal contractor needs to go in setting auditing and monitoring standards as with all safety management arrangements depends on the risk profile of the project. At the lower end of the scale a simple safety inspection using a standard pro forma is adequate; for more complex projects a strategy based on the list above will be required.

The most important issues in being successful with auditing are to have competent, trained auditors, a strategy for identifying what to audit based on risk, and a method of integrating the outputs into the management of the business or project.

Objectives

The fundamental objectives of this section of the plan are to describe to all on site who is auditing what and when. Also how health and safety problems and remedial actions are implemented on site. The section should reinforce the principal contractor's commitment to implementing the standards set in the safety plan.

c. Arrangements for:

(i) Regular liaison between parties on site

Guidance on application

Arrangements for liaison with all parties on site to facilitate the management of health and safety on site should include structured meetings, agendas and the parties. Examples of the types of arrangement this section could cover include:

- project meetings with the client;
- consultants/design coordination team meetings;
- daily site coordination meeting with contractors;
- safety forums with contractors;
- safety report issues to all;
- monthly newsletter;
- health and safety executive via statutory notification;
- emergency planning meetings with emergency services;
- method statement development and briefings;
- reactive safety meetings;
- safety notice board.

The size, nature and complexity of the project should provide a guide to the development of this section. Principal contractors should amend this section as certain contract packages will require more liaison than others. A formal presentation on this section at a pre-start meeting is advisable even for the smallest project.

Objectives

To provide a structured set of arrangements for health and safety to be an integrated element of the project. Also to promote a positive, proactive health and safety culture by providing opportunities for effective liaison between parties.

(ii) Consultation with the workforce

Guidance on application

Consultation with the workforce on a project on site can provide an invaluable source of health and safety-related information from issues regarding occupational health to hazardous site conditions or contractors who could be putting themselves or other workers at risk. Many site workers know about problems on site associated with the environment and work activities and legally and practically must be provided with the opportunity to discuss these issues with the principal contractor. Options for complying with this duty can range from having regular on-site tool-box talks on general health and safety on site to having structured meetings with the contractors or their representatives. In creating a positive health and safety culture on the project, the effective implementation of reasonable recommendations demonstrates to the workforce that they are part of the health and safety team. Consultation with the workforce could be considered in the following circumstances:

- in developing and amending emergency plans and traffic routes;
- in undertaking risk assessments and developing or amending method statements;
- in planning a training session or programme;
- during site health and safety inspections;
- in developing and amending site rules;
- before ordering materials that are hazardous to use or difficult to lift;
- during an accident or near-miss investigation (persons not as witnesses but with site experience to discuss arrangements to prevent reoccurrence).

Objectives

To harness the invaluable experience, site knowledge and cooperation of the workforce as part of the arrangements to manage health and safety on the project. To further promote a positive health and safety culture on a project.

(iii) The exchange of design information between the client, designers, planning supervisor and contractors on site

Guidance on application

This section requires the formal arrangements to clearly define the communication, consultation and approval of design information which is key to any project's success. Design managers, architects and contractors obviously need design information to price, plan and build but also to manage their respective duties under the CDM Regulations. This section in the construction phase health and safety plan could also benefit from defining who a designer is and what a design contribution is, which is not always clear. Typical information in relation to health and safety could include the detailed design of a difficult-to-install component, weights of an awkward or heavy component to assist the contractor in developing a lifting plan, information from designers to contractors on hazards one would not expect a competent contractor to be aware of, a contractor request for information (RFI) from a specific designer, etc.

Arrangements could include:

- a design information register to manage RFIs and the outputs, i.e. inclusion in the method statement, action for the purchaser, inclusion in a lifting plan, consideration or a design amendment by another designer, etc.
- an RFI pro forma;
- the inclusion of a 'health and safety information' column on a design team meeting minutes template;
- a policy on what information is to be provided and what information is not required to avoid unnecessary bureaucracy;
- a policy defining the planning supervisor's method of auditing the health and safety element of the design.

Objectives

To set up arrangements to manage the health and safety element of the design. To take proactive advantage of the team's potential to design and plan for health and safety and to ensure key parties receive safety critical information to facilitate on site health and safety and the development of the health and safety file.

(iv) Handling design changes during the project

Guidance on application

As above, design changes need formal arrangements for approval before being actioned on site. The change will need to be subject to the same scrutiny as all design work, and this section in the plan must avoid changes bypassing the formal system. The planning supervisor must also be provided with the opportunity to assess the health and safety contributions of the designer. A design change can have a significant impact on health and safety, so arrangements for this section would benefit from project gateways to formalise the consultation and sign-off procedure.

Objectives

To control and prevent design changes NOT being subject to the necessary health and safety contributions from the team and supported by designers' health and safety supporting information on residual risk.

(v) The selection and control of contractors

Guidance on application

This section must define who is responsible for the task, which could be an individual or a position within a principal contractor's team, like assistant team leader. The section also must explain the procedure for the following:

- subcontractors;
- designers;
- suppliers who design;
- the self-employed;
- plant and equipment suppliers.

Arrangements must cover or reference the type of questionnaire or interview agenda and the method of appraisal.

In respect of controlling contractors, the principal contractor on any project must have robust procedures with gateways. The following is a list of control issues that should be considered, depending on the nature of the project, before the contractor starts work or starts a new phase of the work:

- signing on and off site, swipe cards, etc. to prevent unauthorised access to the site or areas of the site;
- receipt and approval of selection information in pursuance of the duty to engage competent and adequately resourced contractors and designers;
- induction training covering the site arrangements for safety and activity hazards;
- provision of relevant health and safety information to the organisation, e.g. a copy or relevant section of or access to the construction phase health and safety plan, or a relevant method statement that may interface or conflict with their activities, etc.;

- receipt and approval of method statements and risk assessments for their activities;
- issuing of any permits to work;
- attending site coordination meetings;
- work added to the live programme;
- approval for work to commence granted by the principal contractor.

Site coordination registers provide principal contractors with the potential to manage these procedures, act as an aide-memoire and demonstrate that a control procedure is in place for the project. On simple, relatively low-risk projects with one or two contractors, a simple method statement approval process is sufficient.

Objectives

To ensure only suitably competent and adequately resourced contractors and designers are engaged on the project. To set standards for health and safety from the onset of the project.

(vi) The exchange of health and safety information between contractors

Guidance on application

Information between contractors must be coordinated as on most sites the activities' interface or access to an area is shared. In planning and managing safety, generally undertaken through the development of method statements and risk assessment exercises, the contractor must consider the impact on other contractors and what safety and management arrangements are needed. This valuable information can be coordinated by the principal contractors on large and complex projects or passed directly to contractors and audited by the principal contractor as part of the arrangements. The exchange of information must be made mandatory and this should form part of the arrangements for auditing health and safety on site. Good practice and these arrangements on site can help a subcontractor to invite other contractors to attend method statement briefings to obtain information for them to cascade to their teams and consider when planning for health and safety.

Objectives

To ensure contractors share valuable health and safety information and proactively promote cooperation between contractors so their work activities do not affect the health and safety of others on site.

i) *Security, site induction and on site training*

Guidance on application

Employers, like principal contractors, are required to provide information, instruction and training and also a safe place of work. This section of the plan requires arrangements that could be addressed in the following manner:

- Security: In assessing what arrangements for security are needed as an output of a risk assessment, the following should be considered:

  - pedestrian access security and health and safety;
  - vehicular access security and health and safety;
  - fencing and hoarding;
  - lighting;
  - CCTV;
  - out-of-hours security arrangements;
  - alarm systems;
  - on-site security to control access to certain areas;
  - compound location and access or authorisation;
  - access for emergency services.

- Site induction: Site induction is the necessary training one needs to safely access the site, work on site and respond to an emergency. However, many principal contractors extend this training to cover environmental and quality issues. Such training obviously needs to be suitable and sufficient and, on a small, relatively low-risk site, it will cover the site rules and location of welfare facilities and highlight some of the typical site hazards. On larger sites the training may also include an introduction to the fire plan, traffic plan, method statement approval procedures, work programme and planning procedures, and design management procedures, i.e. almost a presentation on the construction phase health and safety plan. To ensure the trainee has understood the training and can work safely, a test can be incorporated, with a pass mark. Many principal contractors require contractors and self-employed persons to resit the induction following a health and safety offence on site to reinforce the knowledge.

- On-site training: On-site training should be an output of a risk assessment and/or training needs analysis. Typical training courses given on site can vary from health and safety awareness to competence training and assessment for plant operatives. The types of training and refresher training a principal contractor should consider are:

  - health and safety awareness training and passport scheme training;
  - training on the safety management system;

- plant operator training;
- site manager health and safety training;
- inspector training for scaffolding, excavations, etc.;
- slinger and banksman training;
- first aid training;
- confined spaces training;
- appointed person for lifting operations;
- scaffolding and tower scaffolding erection.

Objectives

Security to a standard to ensure the health and safety of third parties and site personnel and also to control personnel and traffic access.

Site induction to be provided for any new person on site and any person who is found not to comply with the site rules to ensure they have sufficient knowledge to work safely on site without risk to themselves and others affected by their activities.

On site training to ensure compliance with statutory provisions and workers are safe and competent to perform their task on site without risk to themselves and others affected by their activities.

(viii) Welfare facilities and first aid

Guidance on application

In setting standards for the provision of welfare facilities a risk assessment of the work activities and assessment of numbers of facilities required will be necessary. The main standards and issues in planning for welfare facilities on site are:

- Provision of toilets, the number to be based on the number of people on site. The facilities must have associated wash basins with water, soap and a means of drying hands. The facilities can be connected to the mains sewer or have a self-contained tank for waste and one for water.
- Provision of washing facilities in the form of basins large enough to allow users to wash their hands, faces and forearms. All basins should have a clean supply of hot and cold running water and a means of drying.
- Provision of rest facilities for taking breaks that are heated and provided with a table and a means of boiling water and heating food.
- Provision of drinking water.
- Provision for storage and drying clothing and personal protective equipment.

For short-duration work the HSE gives guidance on acceptable standards for principal contractors and contractors.

First aid must be addressed by the provision of competent first-aiders. There must be at least one appointed person if there are fewer than five persons on site and at least one first-aider for 5–50 persons on site. For more than 50 on site, there must be one additional first-aider for every 50 employed.

Arrangements must also cover the provision of sufficient equipment to cope with the numbers of people on site and suitable signage of first aid facilities and contact details.

Objectives

To provide and maintain adequate welfare facilities on site to address the needs of the project. To provide adequate first aid resources to address the site.

(ix) The reporting and investigation of accidents and incidents including near misses

Guidance on application

These arrangements will require all accidents and near misses to be reported to the principal contractor for investigation immediately by the quickest practicable means, but the employers are generally responsible for statutory reporting under the Reporting of Injuries, Diseases and Dangerous Occurrences Regulations 1995 (RIDDOR). The arrangements will also need to cover who, within the principal contractor's team, is responsible for undertaking an immediate investigation to assess the situation to prevent reoccurrence and assess site health and safety compliance. It is in the interests of the principal contractor to make these arrangements and especially the requirement to report near misses.

Arrangements concerning subcontractors on site generally require the following information to be supplied to the principal contractor:

- a copy of the report in the accident book;
- the subcontractor's accident investigation containing any remedial action and supporting witness statements (which should be made compulsory);
- a copy of the RIDDOR report if applicable;
- any revised risk assessments or method statements.

Objectives

To comply with the statutory requirements to report accidents and obtain all information on accidents and near misses as part of measuring the construction phase health and safety plan's performance. To react appropriately to these issues to ensure the site and working practices are safe.

(x) The production and approval of risk assessments and method statements

Guidance on application

The production and approval methodology for risk assessments and method statements is critical to health and safety planning, and these arrangements must set standards to achieve:

- project-specific risk assessments and method statements that are suitable and sufficient to control the risks to all those affected;
- the inclusion of all the necessary organisational, resource and logistical elements to implement, monitor and review the safe systems of work;
- an approval process, managed by the principal contractor, that allows sufficient time for the submission's assessment, comments and revision if necessary, consideration of interfaces and impact on site and presentation to the workforce;
- an appreciation of the procedural gateway that ensures no work starts without the principal contractor's formal approval.

Objectives

To control contractors and set standards for the proactive planning and management of health and safety as an integrated part of the contractor's service. To take advantage of the information to assess the impact of the activities on the site's risk profile.

d. Site rules

Guidance on application

The site rules on any project should be made up of a set of the principal contractor's company's health and safety rules and site-specific rules that are derived from the information in design risk assessments and client's information in the pre-tender health and safety plan. The key for the principal contractor is ensuring that the rules are kept up to date, brought to the attention of all on site and enforced effectively. Site rules can cause problems, e.g. hard hats in hot weather, so any relaxation must be authorised by the principal contractor's site representative.

Objectives

To set and enforce clearly defined health and safety standards to avoid accidents, ill health and losses.

e. Fire and emergency procedures

Guidance on application

Fire and emergency plans for rescue from confined spaces or from scaffolding, for example, will require the following considerations:

- the emergency controller;
- key personnel and emergency telephone numbers;
- alarm systems;
- illustrated plans;
- training and testing of the plan;
- equipment;
- specific arrangements.

Objectives

To design and test suitable emergency plans to secure life in the event of an emergency situation.

3. Arrangements for controlling significant site risks

a. Safety risks:

(i) Services, including temporary electrical installation

Guidance on application

This section is designed to cover the safety standards for existing services and the safety issues associated with supplies for task lighting and on-site equipment and supplies to welfare facilities and on-site offices. The main issues to be addressed in this section are:

- liaison with utilities and others, e.g. the Environment Agency, local authorities, etc.;
- installation standards, e.g. BS 7671, the IEE Wiring Regulations for electrical supplies;
- competent persons to undertake work;
- marking and identification of and mechanical protection from site vehicles;
- testing to appropriate standards.

Objectives

To ensure the site establishment and work activities consider the interface with existing services and competent persons install and test to the required standards.

(ii) Preventing falls

Guidance on application

Falling from height, even low-height falls, accounts for the majority of fatalities. This section must clearly set out the minimum standards for the following:

- undertaking work at ground level to avoid or reduce working at height;
- the safe use of ladders, scaffolding including towers, all types of mobile elevation work platforms (mewps), harnesses and fall arrest systems and trestles;
- access control and control of contractors;
- open excavations;
- roof work;
- openings in floors;
- risers;
- inspections and reports on equipment, e.g. scaffolding, mewps, fall arrest equipment, etc.;
- leading edge protection;
- soft landing systems;
- working over or near water;
- materials management;
- falling objects and public safety;
- emergency rescue;
- safety signage;
- training, awareness, safety monitoring, supervision and competence;
- risk assessments and method statements approval.

Objectives

To ensure this significant risk is managed effectively and all work is planned to avoid or set acceptable control measures to the appropriate standards.

(iii) Work with or near fragile materials

Guidance on application

Safety standards for fragile materials particularly roofing material should be addressed in this section. The following issues need consideration:

- risk assessment to identify the hazards and plan installation, replacement access, etc.;
- asbestos;
- existing services;
- access methodology;
- fall protection;
- emergency rescue;
- safety signage.

Objectives

To ensure this significant risk is managed effectively and all work is planned to avoid or set acceptable control measures to the appropriate standards.

(iv) Control of lifting operations

Guidance on application

Safety standards for all lifting operations should be addressed in this section. The following issues need consideration:

- risk assessment to identify the hazards and plan all lifting operations, etc.;
- development of lifting plan to manage risks and implement safety standards;
- appointed person(s) to manage the lifting operations;
- checks, inspection, thorough tests of plant, etc.;
- formation of a crane team to manage lifts;
- existing services;
- third-party safety;
- ground conditions;
- materials stability;
- structural integrity of loading platforms or structure receiving materials;
- man riding equipment, communication, access control, signalling and rescue;
- weather conditions and wind speeds.

Objectives

To ensure all lifting operations are planned, executed in a safe manner and residual risks managed by competent appointed persons as required by the Lifting Operations and Lifting Equipment Regulations 1998.

(v) Dealing with services (water, electricity and gas)

Guidance on application

This section is designed to cover the safety standards for existing services. The main issues to be addressed in this section are:

- liaison with utilities and others;
- location, identification and marking of buried services;
- safe digging procedures;
- marking and barriers for overhead services to avoid contact;
- supervision and control with permits;
- physical and mechanical support;
- the emergency plan.

Objectives

To ensure the site establishment and work activities consider the interface with existing services and liaise with the appropriate authorities.

(vi) The maintenance of plant and equipment

Guidance on application

This section requires clear arrangements for the maintenance of plant and equipment on site. Most equipment on building sites requires maintenance, and all work equipment falls under the scope of the Provision and Use of Work Equipment Regulations 1998 (PUWER), which address maintenance requirements for equipment to be safe. The main issues and areas the principal contractor should focus on in this section to set standards are:

- requesting copies of maintenance reports or schedules for hired or contractor's equipment before work activities start: these arrangements could be part of the procurement process and also defined as a site rule;
- setting up a maintenance schedule for site plant;
- site facilities and a suitable safe area for undertaking maintenance on site;
- allocating responsibility to a principal contractor's representative;
- linking these arrangements with the inspection for specific statutory inspections and tests, e.g. PAT testing for portable electrical appliances, lifting equipment thorough test and examination, etc.;
- PUWER compliance inspection as part of the health and safety auditing regime;
- competence and arrangements of maintenance personnel;
- procedures for preventing use of defective equipment.

Objectives

To ensure all work equipment meets the requirements of the Provision and Use of Work Equipment Regulations 1998 and that all equipment and plant is maintained to a standard to ensure safe operation by a competent person.

(vii) Poor ground conditions

Guidance on application

Arrangements for ground conditions need to take account of slopes, water, stability and existing buried structures. All these issues should be subject to or derive from a risk assessment, and the arrangements set in the construction phase health and safety plan need to address the following:

- flooding, tidal and wet ground where vehicles and site operatives could become trapped;
- site access and access and temporary roads and paths for plant and pedestrians on site;
- unstable ground where falling rocks or mud slides are a risk;
- steep slopes and the risk of plant overturning;
- excavation support;
- voids, buried tanks and wells;
- temporary retaining walls and sheet piling;
- geotechnical surveys.

Objectives

To set standards to address the health and safety problems associated with hazardous ground conditions.

(viii) Traffic routes and segregation of vehicles and pedestrians

Guidance on application

Arrangements for managing traffic and pedestrians on site need to apply the principles of prevention and wherever possible segregate site workers from mobile plant by high-visibility fencing as a minimum standard. The site rules also play an important part in managing these risks with speed limits and competent authorised persons only using and driving plant on site. The main areas of the plan this section needs to focus on are:

- an illustrated traffic management plan to communicate the pedestrian and traffic routes, site access and egress points, lay-down areas, parking facilities, etc.;
- deliveries;

- one-way traffic systems to avoid reversing;
- speed limits and hazard warning signs;
- crossing points;
- access control;
- blind spots;
- training banksmen;
- limits on vehicle size, height and weight;
- overhead services and structures.

Objectives

To establish a set of safety standards to promote the safe communication of pedestrians and site traffic avoiding an interface between the two where practicable. Also to define crossing points and safety rules to manage the risks where there is an interface.

(ix) Storage of hazardous materials

Guidance on application

The standards for the storage of hazardous materials should be derived from risk assessments and the associated hazard data sheets. Also environmental consideration could be addressed here in respect of spillage control. The main issues the principal contractor should consider when planning for storage of hazardous substances are:

- risk assessment;
- restricted or authorised access and security;
- segregated defined storage areas;
- safety equipment and personal protective equipment;
- fire safety arrangements;
- safety signs;
- emergency plans;
- training and information for users, first-aiders and handlers;
- additional first aid arrangements.

Objectives

To provide facilities and the safety arrangements to manage the storage of hazardous substances on site.

(x) Dealing with existing unstable structures

Guidance on application

In respect of setting standards for existing unstable structures the principal contractors will need to cover in this section the arrangements for temporary works, the competence of the temporary works designer and coordinator and site safety issues associated with temporary works. The main considerations the principal contractor must address in managing risks associated with unstable existing structures are:

- obtaining a structural survey to plan either permanent or temporary structural works;
- engaging the services of a competent structural engineer or designer;
- engaging the services of a competent temporary works engineer;
- public safety, access control and security arrangements to the site;
- associated services;
- falling objects and structural collapse as significant hazards;
- the structure's status, i.e. whether the building is occupied;
- emergency planning;
- protection and control of contractors via a permit procedure;
- risk assessment and method statements.

Objectives

To ensure the appropriate actions are taken to make the working and public environments safe and to ensure the works on the structure are planned by competent persons in respect of the design and construction.

(xi) Accommodating adjacent land use

Guidance on application

When sites have no space for the site accommodation, welfare facilities and/or materials storage, the principal contractor will often use adjacent land or facilities. These facilities will require health and safety arrangements that, owing to their location and contractual conditions, may require additional health and safety arrangements to manage the risks. The main issues the principal contractor should consider are:

- access and egress interfaces with highways, client/occupier activities, etc. and security;
- transporting material from a lay-down area off site;
- temporary and existing services;
- pedestrian traffic segregation and interface with highways traffic;
- emergency communication and response.

Objectives

To ensure that suitable health and safety standards are set and maintained on adjacent land used on the project.

(xii) Other significant safety risks

Guidance on application

This section is designed to cover additional safety risks, e.g.

- working over or near water;
- working on highways;
- working in confined spaces;
- roof works;
- erecting structures;
- ground works;
- demolition.

b. Health risks:

(i) Removal of asbestos

Guidance on application

Asbestos arrangements will need to cover identified asbestos as part of a survey from a client or in a pre-tender health and safety plan and reactively when identified by site personnel. The arrangement will need to address the competence of operatives, control of contractors, the working standards, notification to the HSE, waste management and third-party health and safety. Reference in this section could be made to a specific set of asbestos health and safety rules and awareness training. The principal contractor or contractor could also, as an output of a risk assessment, consider the programming of the works to avoid interface with the other contractors or the client, e.g. removing asbestos in a school during school holidays. Difficult-to-manage issues with asbestos standards on site concern when an HSE-licensed removal contractor is needed to remove the material. HSE guidance should be used to ensure licensed contractors are used for all removal except work on asbestos cement sheet, or asbestos cement products like duct- and pipework; however, the work must be done in compliance with the Control of Asbestos at Work Regulations 2002.

Objectives

To ensure minimum heath and safety standards in relation to asbestos on site and avoid any exposure to site workers, members of the public and future building users.

(ii) Dealing with contaminated land

Guidance on application

The standards for working safely on a contaminated land site must control contamination and contain it to the site and transfer site and must address hygiene and occupational health issues. The risks can vary from relatively low-risk inert waste materials to highly dangerous chemicals. The site rules, induction, training and access control will be key areas in managing the risk. To control contamination, vehicle wheel wash stations, hygiene facilities for site workers, suitable personal protective equipment and enforced safety rules regarding eating and working practices will need to be addressed in this section. Awareness and induction training will need to cover the hazards and controls, and auditing frequencies will need to be set to monitor the standards.

Objectives

To ensure the workers on site, the general public and anyone accessing the site are aware of the risks, control measures and emergency procedures.

(iii) Manual handling

Guidance on application

The risks associated with manual handling on site will require a strict set of conditions in this section, enforced by the principal contractor and sub-contractors. The conditions can be company specific and produced as a result of a manual handling risk assessment. The main issues the principal contractor will need to cover in this section are:

- the minimum acceptable weights to lift on site for one-man and two-man lifts to provide operatives with a solid rule-of-thumb guide;
- training on manual handling hazards and lifting techniques;
- provision of lifting equipment and guidance on its application either through training or via safety posters;
- manual handling risk assessment;
- occupational health arrangements to monitor personnel and advise.

Objectives

To ensure no site operative suffers from an injury as a result of inappropriate manual handling.

(iv) Use of hazardous substances

Guidance on application

This section will need to cover the risk assessment requirements and minimum standards for products that contain hazardous substances and by-products of construction operations that contain hazardous substances. The standards for personal protective equipment, disposal, storage and emergency procedures including first aid will also need to be derived from risk assessments by the respective contractors. Hazardous substances registers may need to be developed to ensure that a suitable emergency or first aid response is effective. Also the principal contractor will need this information to undertake appropriate audits to evaluate the contractor's risk assessment, selection and use of control measures and effects on other site workers possibly affected by any substances.

Objectives

To ensure that all potentially hazardous substances are subject to a risk assessment and where products or activities cannot be reasonably prevented suitable control measures are developed and employed on site to control exposure to acceptable levels.

(v) Reducing noise and vibration

Guidance on application

The principal contractor will require in this section arrangements for contractors to undertake risk assessments of activities involving noisy equipment and equipment that has the potential to cause occupational health problems associated with exposure to vibration. The assessments will be based on previous knowledge, information supplied with equipment and monitoring results on site. The competence of the assessor will be defined in this section with the requirement to periodically monitor the activities. Principal contractors and contractors will also need to share this health and safety information. Noise operations can cause problems of not being able to hear alarms or safety instructions. Again planning and coordination of contractors can provide opportunities for certain activities that cause potential problems to other workers or affect the client's operations to be undertaken out of normal working hours. It is advisable for principal contractors to have their own noise and vibration monitoring arrangements to monitor levels and occupational health arrangements to monitor contractors' health and safety performance.

Objectives

To ensure that noise and vibration hazards are identified and assessed and where they cannot be reasonably avoided adequate control measures developed and implemented on site to protect operatives and others affected by their activities.

(vi) Other significant health risks

Guidance on application

Other significant health hazards that one could provide arrangements for in detail in this section are:

- lead paint;
- lasers;
- radiation;
- leptospirosis (Weil's disease);
- whole-body vibration;
- anthrax;
- zoonoses;
- legionellosis;
- excessive heat;
- arsenic in paint.

4. The health and safety file

a. Layout and format

Guidance on application

The layout or structure of the health and safety file should be defined in the pre-tender health and safety plan. If this is not the case, arrangements should be detailed here on liaison with the relevant parties, generally the client's representative, to plan the file's structure and also the format, e.g. electronic, hard copy, both, web based, numbers of copies, etc.

Objectives

To ensure a user friendly health and safety file is developed for use during the project and for providing information in managing health and safety on future projects and maintenance work.

b. Arrangements for the collection and gathering of information

Guidance on application

This section must identify the key personnel and arrangements for obtaining, approving and collating the file information. These arrangements depend on the planning supervisor's service level agreement. On some projects the principal contractor develops the file and the planning supervisor monitors its development. On generally smaller projects, the information is sent to the planning supervisor for approval, collation and handover to the client. One of the important issues in this section is the contractual arrangement for contractors to provide the information to the principal contractor.

Objectives

To ensure all contractors and suppliers supply appropriate health and safety information for the development of the health and safety file as a part of their contract and services.

The health and safety plan is designed to be a live document, which should continue to evolve throughout the construction phase and must reflect any changes in the management or health and safety arrangements for the project. To this end an amendments section in the initial part of the document provides an opportunity to manage and inform users of the changes.

There has been evidence, since the introduction of the regulations, of non-project-specific, almost generic, plans being developed with insufficient thought given to the significant hazards associated with the scheme. In defence of the principal contractor, if they are starved of project-specific information from the planning supervisor the development of the construction phase health and safety plan may feel rather more like paper management than a hazard management exercise.

Once the principal contractor believes that the plan has sufficient information for work to begin on site, the document must be sent to the client for assessment or to the planning supervisor, who may be responsible for advising the client on the plan's suitability. Only when confirmation has been received from the client, client's agent or perhaps the planning supervisor on behalf of the client that the construction phase health and safety plan is acceptable may the construction phase start.

Ensuring cooperation between contractors

The control of contractors is one of the primary functions of the principal contractor, and ensuring they cooperate on site and when planning for health and safety is essential. The requirement to cooperate with other contractors should, where practicable, start before work on site starts. As part of the risk assessment process and in developing method statements, contractors must consider the impact on other persons on site, and issues like access, deliveries, lifting operations, production of dusts, demolition, etc. all impact on other contractors' health and safety.

So the principal contractor must set up the appropriate arrangements to ensure cooperation. These arrangements should form part of the construction phase health and safety plan, which in turn can form part of the contractor's form of contract. The arrangements generally involve attending site coordination meetings, sharing risk assessments and method statements and liaising on site as part of the day-to-day managing of health and safety.

(1) The principal contractor appointed for any project shall –

(a) take reasonable steps to ensure co-operation between all contractors (where they are sharing the construction site for the purposes of regulation 11 of the Management of Health and Safety at Work Regulations 1999 or otherwise) so far as is necessary to enable each of those contractors to comply with the requirements and prohibitions imposed on him by or under the relevant statutory provisions relating to the construction work.

Extract from CDM regulation 16

If the principal contractor is to accurately coordinate and manage the health and safety aspects of the project then the interrelationship between the respective contractors' activities must be fully and proactively assessed from this holistic position as part of the project planning to avoid conflicts and access problems on site. One would hope that information on potential pinch points or restricted access, if associated with a difficult-to-manage construction activity, would be provided to the principal contractor via the designer's risk assessment or pre-tender health and safety plan.

In summary, principal contractors must set up and enforce arrangements to promote cooperation between contractors and the sharing of relevant health and safety information. To achieve this, a coordination role must be played by the site, construction or project manager.

Controlling site access

The principal contractor needs to ensure that only authorised people access the construction site and are aware of the site rules, hazards and emergency procedures as a minimum. Conversely, the principal contractor must ensure that no unauthorised persons gain access to the site or certain areas of the site. This public safety duty, it could be argued, is one of the most difficult challenges and also the most significant business risk issue, given the social inner city problems with young children prepared to take risks and trespass on to building sites for recreation.

> (1) The principal contractor appointed for any project shall –
>
> (c) take reasonable steps to ensure that only authorised persons are allowed into any premises or part of premises where construction work is being carried out.
>
> Extract from CDM regulation 16

> Breach of a duty imposed by these Regulations, other than those imposed by regulation 10 and regulation 16(1)(c), shall not confer a right of action in civil proceedings.
>
> CDM regulation 21

To emphasise the significance of this duty, if a principal contractor is found to have breached this regulation the path is open for civil action against them as there is no exclusion from civil liability.

The principal contractor will therefore be required to introduce authorisation procedures for all workers and site visitors. The extent of control over site access will be based on an assessment of risk, considering such issues as the site location, the surrounding environment and the type of construction activity being undertaken.

Suitable arrangements are then necessary to ensure that unauthorised cohorts do not gain access to the site. The ACoP suggests that all sites should be physically defined, where practical, by suitable barriers. It also mentions that additional consideration should be given if there is evidence of children playing on, or near, the site.

472 children killed or injured on construction sites in a five-year period.
Source: Fatalities and injuries to children (aged 1–15) in the construction sector, as reported to the HSE during the period 1998/09 to 2002/03.

Safety signs, signing-in and -out books, stickers on hard hats indicating that induction training and authorisation to the site have been given, access control swipe cards and security guards are common arrangements on projects.

Displaying the project notification

It is the responsibility of the principal contractor to clearly display on the site, in a prominent position, the project information notified to the Health and Safety Executive. More often than not the principal contractor will make a copy of the original notification sent to the HSE (F10) and attach it where it can be easily read by those working on the site or indeed affected by the work.

Enforcement of the site rules

The principal contractor, as the duty holder ultimately responsible for health and safety on site, has a duty to ensure all site rules are adhered to.

> (1) The principal contractor appointed for any project shall –
>
>> (b) ensure, so far as is reasonably practicable, that every contractor, and every employee at work in connection with the project complies with any rules contained in the health and safety plan.
>
> Extract from CDM regulation 16

The 'rules', which may come from the client as well as the principal contractor, must be communicated to all relevant cohorts, ideally through the plan itself, site inductions, appropriate signage and toolbox talks. Procedures for non-compliance with any rules also need to be made clear. Site inspections and monitoring will generally assess the levels of compliance.

Below is an example of a building project's rules extracted from the construction phase health and safety plan:

Workforce

- Make sure you have attended a site induction prior to commencing work.
- Safety helmets, high-visibility vests and safety footwear are mandatory on site.
- Make sure your supervisor has briefed you prior to starting work.
- Wear the personal protective equipment specified at all times.
- Do not operate plant or equipment without authorisation or correct PPE.
- If you have to use glues, oils, fuels, etc., ensure you have been briefed.
- No alcohol, drugs, radios or walkmans allowed on site.
- Only eat, drink or smoke in areas designated by management.
- Look out for your fellow workers and they will look out for you.

Work areas

- Do not cross barriers or enter no-go zones.
- Keep your work area tidy and put rubbish in piles or bins away from access routes.
- Keep all accesses clear of debris, materials and cables.
- Holes through floors MUST be covered, marked and adequately fixed.
- Some areas on site will require permits to access. These areas are clearly defined.

Permits

- Some activities need special permits; if you haven't received a briefing, ask.
- With hot work permits, know which extinguisher to use for different types of fire.

Plant and equipment

- Scaffolding is only to be erected, dismantled or altered by competent scaffolders.
- Mobile towers can only be erected or modified by approved erectors.
- Only appropriate certificated operators to drive or operate plant (including disc cutters).
- Electrical tools have date-tags to keep the machine safe; don't remove them.
- Do not pour old oil or other pollutants down the drain.

Unplanned events

- Report all accidents to the principal contractor so we can learn from our mistakes and prevent recurrence.
- Know the best means of escape from your work area, and your assembly point.
- Fight fires quickly with the correct extinguisher if competent, after raising the alarm.
- We will not tolerate damage to company property, safety equipment or personal property, but we cannot be responsible for your own possessions or cars.
- Unsafe working will not be tolerated and positive action will be taken.

General

- Take notice of signs and instructions. Do not bring children or pets to site.
- Inform your supervisor if you think something is unsafe.
- Keep noise and dust levels down to the minimum and wear your PPE.

Incentive schemes that promote compliance with the site rules can help in a positive way, but effective enforcement generally requires disciplinary action and, in the worst cases, contractors or companies may be required to resit induction training, be removed from site or be asked to stop work until they can demonstrate that remedial action has been taken to avoid re-offending.

Provision of relevant information to contractors

It is important for all contractors working on a project to be provided with relevant information to help them plan for health and safety. The principal contractor needs to provide such information.

> The principal contractor appointed for any project shall ensure, so far as is reasonably practicable, that every contractor is provided with comprehensive information on the risks to the health and safety of that contractor or of any employees or other persons under the control of that contractor arising out of or in connection with the construction work.
>
> CDM regulation 17(1)

Contractors are expected to provide risk assessments and associated method statements and are unable to effectively carry out this duty without project-specific information. The principal contractor must ensure that they only provide information that is relevant to the contractor's specific work area or shared area, activity or environment. The provision of irrelevant information for a contractor may obscure significant risks. The principal contractor's arrangement in the construction phase health and safety plan must identify who is responsible for providing this information.

Listening to the workforce and their representatives

As well as being obviously good practice it is also a legal duty for the principal contractor to engage the workforce in terms of discussing health, safety and welfare issues. Many accident and near-miss investigations have identified that site workers were aware of the hazardous environment or unsafe working practice prior to the specific incident. Workers' experience and input must be taken into account.

> The principal contractor shall –
>
> (a) ensure that employees and self-employed persons at work on the construction work are able to discuss, and offer advice to him on, matters connected with the project which it can reasonably be foreseen will affect their health and safety; and
>
> (b) ensure that there are arrangements for the co-ordination of the views of employees at work on construction work, or of their representatives, where necessary for reasons of health and safety having regard to the nature of the construction work and the size of the premises where the construction work is carried out.
>
> CDM regulation 18

This two-way dialogue is an important contribution towards successful health and safety management on site. The combined experiences of persons working on site are a valuable asset and it is in everybody's interest to have this exchange of ideas.

The principal contractor should instigate site safety meetings with all relevant parties, discuss hazardous operations and safe systems of work, and actively encourage dialogue.

Health and safety training and information

The legal duty and objective of the principal contractor in respect of training and information should be based on risk. Training in the form of induction training and awareness training is to ensure contractors are aware of the risks to their health and safety, and the rules and arrangements to manage and control the risks. Induction training should generally cover the site rules, significant hazards, emergency procedures and security and site access arrangements.

The information to be issued to contractors is highlighted in the CDM Approved Code of Practice, which states:

> Good communication is essential to co-operation and risk control. Information about risks and precautions can be communicated by, for example:
>
> (a) drawings that highlight hazards or unusual work sequences identified by designers, with clear advice on where to find more information;
> (b) the relevant parts of the plan;
> (c) meetings to plan and co-ordinate the work;
> (d) effective arrangements to discuss the plan with those involved;
> (e) making the plan available to workers and their representatives;
> (f) induction training and toolbox talks to ensure workers understand the risks and precautions;
> (g) providing a leaflet explaining the site rules that can be given to everyone at the induction training.

Formal procedures to record the receipt of information and briefings and even the testing of induction training safety information can help focus the contractor's attention on the subject, especially if they appreciate that it is linked with authorisation to access the site.

The health and safety file

The principal contractor is charged with the responsibility to collect the health and safety file information from the respective subcontractors, designers and suppliers. The arrangements for the collection of the information, which can be clearly defined in the contractors' contracts, must be entered in the construction phase health and safety plan and should identify the responsible person. Arrangements can make this management process part of the project progress meetings or contractor coordination meetings to formalise the requests for overdue and unsuitable information.

Some projects using portal web-based information management systems can be set up to receive this information throughout the project, which provides a live environment for the planning supervisor to monitor the development of the file and comment well before the end of the project.

The structure of the file is generally defined in the pre-tender health and safety plan to suit the clients and end users.

Managing problems

Q. *Can there ever be more than one principal contractor on a project?*

A. No. There can be two main contractors, but the CDM Regulations have been designed to have a controlling health and safety manager with overall responsibility and authority on a project.

Q. *Is the principal contractor always the main contractor on a project?*

A. In many cases the main contractor will assume the role of principal contractor – but not always. The regulations demand only that the principal contractor be a contractor and also be competent and adequately resourced. There are examples of clients appointing themselves to the role, project management teams assuming the duties, and even subcontractors acting as principal contractor.

Q. *What options does the principal contractor have if the information provided in the pre-tender plan is inadequate?*

A. This is a common problem and places the contractor in a difficult position. Information on the client's activities, site access, and difficult-to-manage design or environmental risks, e.g. asbestos or buried services, can be very helpful. Experienced contractors will as part of their risk assessment exercise, which can start in the development of a tender response, start developing a schedule of requests for information (RFIs) to obtain surveys, client information on existing activities on site, etc., and many tendering processes allow the contractor the opportunity to visit the site and discuss the project with a representative, which in turn provides the opportunity to ask questions and obtain information on hazards.

Q. *Is the principal contractor also a designer, on a design-and-build contract for example?*

A. If, on a design-and-build-type contract, the principal contractor is contributing to the design of the structure, then CDM regulation 13 duties will apply to them.

Q. *How much detail should the principal contractor require for the construction phase health and safety plan?*

A. The detail should be proportional to the health and safety risk profile of the project. The CDM ACoP provides a list of what the minimum contents of a construction phase plan should contain, which are:

    (a) general information about the project, including a brief description and details of the programme;
    (b) specific procedures and arrangements for the early work;
    (c) general procedures and arrangements which apply to the whole construction phase, including those for the management and monitoring of health and safety;
    (d) welfare arrangements;
    (e) emergency procedures; and
    (f) arrangements for communication.

Q. *Does the principal contractor need to request risk assessments from all contractors?*

A. The contractors have a duty to cooperate with the principal contractor, and this duty requires them to provide health and safety information. The principal contractor can request the risk assessment as part of their duty to ensure contractors cooperate and as part of their duty to provide health and safety information.

Q. *If a client's existing operations or working practices change and impact on the principal contractor's construction site what action should one take?*

A. The principal contractor's duty to monitor health and safety on site and undertake a risk assessment provides the legal agenda to react to the changes and revise the health and safety arrangements as necessary. Principal contractors in control of sites generally and correctly treat the clients as contractors, which will require the submission of risk assessments and method statements within a pre-defined timescale before the intended start of work for approval. This information enables the principal contractor to consider the impact on site holistically and to provide all their contractors with the necessary information so they can consider the impact on their operations too.

Chapter 6

The contractor

Introduction

As discussed in Chapter 1, the construction industry is both the largest industry in the UK and one of the most dangerous. On average over two people have died every week on construction sites and hundreds have suffered major injuries and ill health over the last few decades. The majority of these people are clearly contractors.

> '[C]ontractor' means any person who carries on a trade, business or other undertaking (whether for profit or not) in connection with which he –
>
> (a) undertakes to or does carry out or manage construction work,
> (b) arranges for any person at work under his control (including, where he is an employer, any employee of his) to carry out or manage construction work.
>
> Extract from CDM regulation 2(1)

If CDM is to succeed in protecting people at work in the construction industry then contractors are amongst the very people it is aimed at protecting through solid design, planning and foresight. Contractors themselves must obviously contribute to the health and safety management process and have several significant duties under CDM itself.

Before we examine the CDM contributions expected from contractors it is worth reminding ourselves of the statutory duty imposed on them as employers in section 2 of the Health and Safety at Work etc. Act 1974, which states that:

> It shall be the duty of every employer to ensure, so far as is reasonably practicable, the health, safety and welfare at work of all his employees.
>
> HSWA section 2(1)

Section 3 of the Act also emphasises the duty on an employer to conduct his work in such a manner, as far as is reasonably practicable, as to ensure that his activities do not cause harm to people not in his employment. This is generally aimed at protecting the general public and other non-employees. Significantly these section 3 duties also apply to self-employed contractors, who are widely found throughout the construction industry.

> Every self-employed person must conduct his undertaking in such a way as to ensure that persons not in his employment are not exposed to health and safety risks.
>
> HSWA section 3(2)

Section 2 and section 3 of the Act are supported by regulation 3 of the Management of Health and Safety at Work Regulations, which offers:

Every employer shall make a suitable and sufficient assessment of:

(a) the risks to the health and safety of his employees to which they are exposed whilst they are at work, and

(b) the risks to the health and safety of persons not in his employment arising out of or in connection with the conduct by him or his undertaking.

MHSW regulation 3

Other significant legislation applicable to contractors provides a framework for the health and safety management tools they use. Examples are shown in Table 6.1.

Contractors are CDM duty holders with a set of regulations to promote their interaction with the principal contractor and manage health and safety. The extent to which a contractor must go to execute these duties generally is defined in the project's construction phase health and safety plan, which must be communicated to the contractor.

The contractor's duties explained

Now that we have an appreciation of the other health and safety duties imposed on contractors (employers and self-employed), we will look in detail at the CDM requirements placed on them. The contractor's duties and general CDM duties are summarised as follows:

- appointing competent and adequately resourced contractors or designers;
- cooperation with the principal contractor;
- provision of information to the principal contractor;
- complying with directions from the principal contractor;

Table 6.1 Examples of legislation and associated health and safety tools

Legislation	Health and safety tools
Construction (Health, Safety and Welfare) Regulations 1996	Site health and safety checklist Inspections and examinations register
Work at Height Regulations 2005	Working at height risk assessment template
Confined Space Regulations 1997	Permit to work/enter
Lifting Operations and Lifting Equipment Regulations 1998	Lifting plan template Lifting equipment checklist
Management of Health and Safety at Work Regulations 1999	Risk assessment pro forma Method statement template Traffic management plan Induction training programmes Toolbox talk programme

- complying with relevant rules from the health and safety plan;
- reporting accidents and incidents to the principal contractor;
- receiving information prior to starting work.

Appointing competent and adequately resourced contractors or designers

Many contractors will obviously appoint other contractors and designers themselves. When this is the case it is a requirement to ensure that they are both competent and adequately resourced to undertake the work. The contractor making the appointments will consequently need to comply with relevant duties imposed by CDM regulations 8 and 9, namely:

(2)   No person shall arrange for a designer to prepare a design unless he is reasonably satisfied that the designer has the competence to prepare that design.

(3)   No person shall arrange for a contractor to carry out or manage construction work unless he is reasonably satisfied that the contractor has the competence to carry out or, as the case may be, manage, that construction work.

Extract from CDM regulation 8

(2)   No person shall arrange for a designer to prepare a design unless he is reasonably satisfied that the designer has allocated or, as appropriate, will allocate adequate resources to enable the designer to comply with regulation 13.

(3)   No person shall arrange for a contractor to carry out or manage construction work unless he is reasonably satisfied that the contractor has allocated or, as appropriate, will allocate adequate resources to enable the contractor to comply with the requirements and prohibitions imposed on him by or under the relevant statutory provisions.

Extract from CDM regulation 9

The arrangements for compliance will need to demonstrate a process of assessing designers and contractors by obtaining information through either a questionnaire or an interview, or by both methods. The information will need to be subject to an appraisal by a competent assessor, i.e. someone with the experience and skill to make a judgement on their competence and resources. The balanced approach to this exercise is to avoid any irrelevant questions and to focus the questions on what is appropriate based on the risk profile of the project. A painting company internally painting some flats does not require the same assessment as a steel frame company erecting a tower in the middle of a city. Likewise the information received should be proportionate. Standards questionnaires can promote bureaucracy, and a statement at the top of the questionnaire requesting the respondent to provide only relevant information to the relevant questions or recommending to the respondent that they are not required to answer all questions can help. Having a select list of those whose competence has been established previously is also efficient, and only resources will need assessment for a particular project.

Cooperation with the principal contractor

The relationship between contractors and the principal contractor is fundamental to successful hazard management on site. Cooperation in relation to planning and programming work on site to reduce risk is a fundamental part of this duty. As part of the contractor's health and safety management process generally achieved through risk assessment, the contractor can liaise with the principal contractor to assist them in managing health and safety on site. A simple example is attending a site coordination meeting to plan the programme of which contractors can access which part of the site or to plan the assembly of a crane.

> Every contractor shall, in relation to the project –
>
> (a) co-operate with the principal contractor so far as is necessary to enable each of them to comply with his duties under the relevant statutory provisions.
>
> Extract from CDM regulation 19(1)

Principal contractors must have this cooperation if they are to comply with their own CDM responsibilities to ensure cooperation between contractors and to ensure that every contractor on the project complies with the rules laid down in the construction phase health and safety plan and the provision of information to the planning supervisor.

Provision of information to the principal contractor

The contractor is required to provide the principal contractor, as soon as possible, with any information which may affect the health and safety of site workers or indeed others who may be affected by their activities such as the general public. The contractor shall:

> (b) so far as is reasonably practicable, promptly provide the principal contractor with any information (including any relevant part of any risk assessment in his possession or control made by virtue of the Management of Health and Safety at Work Regulations 1999) which might affect the health and safety of any person at work carrying out the construction work or of any person who may be affected by the work of such a

person at work or which might justify a review of the health and safety plan.

Extract from CDM regulation 19(1)

As discussed earlier, risk assessments will need to be carried out by contractors to satisfy their duty under regulation 3 of the Management of Health and Safety at Work Regulations 1999. These may well form part of their strategy for complying with this responsibility of providing information to the principal contractor.

This duty also requires the contractor to provide information in respect of the health and safety file. Such information could include as built or installed drawings, operational and maintenance manuals, method statements for more complicated procedures, contact details of specialist suppliers and information on residual risks for any future maintenance, cleaning and refurbishment work.

Example: A contractor planning the demolition of a wall on site recommended a site meeting with the principal contractor to plan a revised traffic route. The contractor provided the principal contractor with an illustrated method statement detailing the exclusion zone and traffic management arrangements for the zone. The principal contractor used the information to plan a revised traffic management route and communicated the issues to the site team at a coordination meeting.

Complying with directions from the principal contractor

To enable the principal contractor to manage health and safety regulations and control contractors, they are given duties which are supported by the duty of contractors to cooperate and comply with the directions of the principal contractor. To this end it is a CDM offence for a contractor not to comply with the directions of the principal contractor to secure health and safety. The contractor shall:

(c) comply with any directions of the principal contractor given to him under regulation 16(2)(a).

Extract from CDM regulation 19(1)

Assuming the directions are reasonable and necessary for the principal contractor to comply with their duties under CDM, the contractor will need to comply. However, the duty to cooperate should, if executed effectively, avoid the need for the principal contractor to enforce this duty. These directions can be directly related to health and safety action or can be higher-level programming and planning issues to facilitate health and safety on site.

Complying with relevant rules from the health and safety plan

It is clearly in the interest of the contractor to comply with all relevant site rules in terms of protecting the health, safety and welfare of their employees and others affected by their work. These rules and standards obviously need to be presented to the contractors so that they are aware of the responsibilities which must be addressed when developing risk assessments, method statements, lifting plans, etc.

Contractors will need to take steps to ensure that they comply with this regulation. This is often undertaken by contractors inspecting and monitoring their own work to ensure that their employees or indeed self-employed workers they are supervising are following the rules and standards set by the principal contractor in the health and safety plan. The Approved Code of Practice goes a stage further and offers that:

> Such monitoring may identify shortcomings in the plan. Where this is the case, the contractor should ensure that the principal contractor is told.
>
> CDM Approved Code of Practice

Contractors will also need access to the construction phase plan's updates and revisions to comply.

Reporting accidents and incidents to the principal contractor

If the contractor has to notify the Health and Safety Executive of an accident or incident as required by the Reporting of Injuries, Diseases and Dangerous Occurrences Regulations (RIDDOR), CDM also provides them with an additional duty to promptly inform the principal contractor. As the principal contractor is responsible for health and safety throughout the construction phase then this information is required to monitor compliance levels with relevant health and safety legislation. Also, the proactive principal contractor will review existing health and safety management arrangements to ensure they are effective. These issues should be covered in induction and form part of the site rules.

Receiving information prior to starting work

The provision of information to the contractor is the end of the road in terms of filtering residual hazards through the project for the construction phase. Regulation 19(2), (3) and (4) places the onus on the employer or indeed self-employed contractor to ensure that they and their employees receive appropriate health and safety information.

(2) No employer shall cause or permit any employee of his to work on construction work unless the employer has been provided with the information mentioned in paragraph (4).

(3) No self-employed person shall work on construction work unless he has been provided with the information mentioned in paragraph (4).

(4) The information referred to in paragraphs (2) and (3) is –

(a) the name of the planning supervisor for the project;

(b) the name of the principal contractor for the project; and

(c) the contents of the health and safety plan or such part of it as is relevant to the construction work which any such employee or, as the case may be, which the self-employed person, is to carry out.

Extract from CDM regulation 19

Remember that the name of both the planning supervisor and the principal contractor will be displayed on the notice required under regulation 7. It is for the contractor though to provide the information to employees and the self-employed.

The respective elements from the health and safety plan must also be articulated to the employees and self-employed by the contractor. The contractor will receive, from the principal contractor, comprehensive information on the specific health and safety risks, generally in the form of the construction phase health and safety plan. This must be assessed and communicated effectively. Depending on the specific risks, this can be achieved, for example, by toolbox talks, method statement briefings and induction and awareness training.

The principle of the contractor adopting an informative and inclusive policy in terms of their employees or the self-employed when managing health and safety is a sensible regime. The relevant experiences of the contracting team will no doubt contribute towards managing the risks.

Some frequently asked questions

Q. *I am a subcontractor on a project. Do I have contractor duties under the CDM Regulations?*

A. Yes. The CDM Regulations define a contractor as:

any person who carries on a trade, business or other undertaking (whether for profit or not) in connection with which he –

(a) undertakes to or does carry out or manage construction work,

(b) arranges for any person at work under his control (including, where he is an employer, any employee of his) to carry out or manage construction work.

Extract from CDM regulation 2(1)

Although contractors are often subcontractors in that they are engaged by another contractor or by the client to work with other contractors, the term 'subcontractor' has no legal connotation as far as CDM is concerned.

Q. *I am having problems accessing an area on site safely because of the number of different contractors in that area. What action should I take?*

A. Both legally and practically, one should provide the principal contractor with a risk assessment highlighting the interface and level of risk. These issues should be coordinated by the principal contractor at coordination meetings, the output of which should be a realistic construction programme.

Q. *I have been asked to take over a small site as principal contractor because the main contractor who was the principal contractor has finished the contract. What should I do?*

A. There are competence, resourcing and insurance issues here, and the best advice is to consult a construction safety consultant or indeed the Health and Safety Executive. The issues to consider are:

- the risk profile of the project: a small, low-risk project or phase of the project, e.g. painting and decorating an empty building with a low number of employees, will generally require limited competence and resources;
- the management systems and resources to comply, which will include a change in contract.

Chapter 7

CDM management toolkit

CDM policy ref.		Issue	Date
Client section			
C1	Client project CDM checklist		
C2	Client's letter of appointment to HSE		
C3	Planning supervisor's prequalification		
C4	Designer's prequalification		
C5	Principal contractor's prequalification		
C6	Health and safety information survey pro forma		
C7	Certificate of health and safety file handover		
C8	Planning supervisor's questionnaire appraisal		
C9	Client information to planning supervisor checklist		
Designer section			
D1	Designer CDM checklist		
D2	Design risk assessment toolkit		
Planning supervisor section			
PS1	Planning supervisor's CDM checklist		

CDM policy ref.		Issue	Date
Planning supervisor section (continued)			
PS2	Designer's questionnaire appraisal		
PS3	Principal contractor's questionnaire appraisal		
PS4	Construction phase plan assessment		
PS5	Authorisation for the commencement of work		
PS6	Project risk register template		
Principal contractor section			
PC1	Principal contractor CDM checklist		
PC2	Contractor's prequalification		
PC3	Contractor's questionnaire feedback appraisal		
PC4	Construction phase plan contents checklist		
PC5	Site safety coordination register		
AP1	CDM audit pack		

CLIENT PROJECT CDM CHECKLIST

Form C1

Project name	
Project ref. no.	

Ref.	Project stage	CDM requirement	Considerations	Completed by	Date
1	Project start	Appoint client agent *(option for client if lacking competence or resources)* Requirement for this project ☐ Yes ☐ No	The designer will inform the client of their CDM responsibilities. The planning supervisor will also be able to offer guidance. *Refer to Form C2*		
2	Feasibility (early design stage)	**Appoint planning supervisor** **Establish competence** **Ensure allocation of resources**	The planning supervisor potentially makes their largest contribution at the initial/ early design stage. Early appointment essential. *Refer to Form C3* *Refer to Form C3*		
3	Feasibility (early design stage)	Establish a service level agreement with the planning supervisor	This is important to clarify the extent of the appointment and define what services the planning supervisor will and will not carry out for the project. *Refer to Form PS2*		
4	Pre-design/ design	**Provision of relevant health and safety information about the existing structure and the site**	The design team requires information to consider the relevant health and safety implications. This must be provided from the earliest design stages. The client can make a significant contribution to the project by providing accurate information in good time. Examples include information on asbestos, contaminated ground, the location of buried and overhead services, structural survey. *Refer to Form C9*		
5	Design	**Appoint designer(s)** – *only if appropriate*	Designers must be able to address the health and safety implications likely to be involved with their designs. The client must ensure this is the case. If appointments are in-house, competence and resources still need to be satisfied. The planning supervisor will be available to advise.		

continued overleaf

#	Phase	Action	Notes
		Establish competence	*Refer to Form C4*
		Ensure allocation of resources	*Refer to Form C4*
6	Design	**Appoint principal contractor**	Regardless of the procurement route or the scale of the project, if the CDM regulations are applicable the client must appoint a principal contractor. They must be both competent and adequately resourced to undertake the project. Competence may be established previously on similar type projects but adequate resources must be established for each project. The planning supervisor will be available to advice.
		Establish competence	*Refer to Form C5*
		Ensure allocation of resources	*Refer to Form C5*
7	Pre-construction	**Ensure that the construction phase health and safety plan is suitably developed for construction work to start**	The planning supervisor will be able to advise on the suitability of the plan at this stage. Note, however, that the legal responsibility remains with the client to satisfy this regulation. **Note that a breach of this regulation carries no exclusion of civil liability. See CDM regulation.** *Refer to Form PS5*
8	Post-construction	**Confirm that the health and safety file has been received**	It is prudent to inform the planning supervisor that the health and safety file has been received.
9	Post-construction and throughout operation and maintenance	**Ensure that the health and safety file is available for inspection and accurately maintained**	Consider that the file may be required at an earlier stage, e.g. partial occupation, phased handover, or maintenance work prior to completion of the project. Note that the file is a valuable source of information for future work and if properly maintained can assist with reducing costs.
10	When ownership changes hands	**Ensure that the health and safety file is handed over**	

As project CDM client's representative I confirm that so far as is reasonably practicable all CDM client/ client's agent duties within my control have been complied with.

CDM sign-off	Date

Project review:

Hazard/ issue	Client management action (*Please comment on how hazards have been successfully managed by the client's CDM and project management actions*)
Summary of other issues	Lack of information from client or other designers to undertake effective risk assessment; problems obtaining suitable information for health and safety file

Copy to safety adviser (for audit or policy development)	Sent to	Date sent

The Health and Safety Executive
Postcode

Dear Sir/Madam

Re: *Project* _____

Construction (Design and Management) Regulations 1994
CDM regulation 4 – Declaration regarding transfer of client's duties to a CDM agent

Please accept this letter as a formal declaration to the HSE that _____ are acting as CDM client agents for [*name of client here*] on the above-named project. In respect of these duties please see the attached client's project checklist detailing our arrangements for implementing the company's health and safety policy arrangements in respect of CDM and the client's agent duties.

If you require any further information please do not hesitate to contact the above.

Client's contact details

Company	
Contact	
Address	
Telephone	
Email	
Fax	

Yours sincerely,

Attachment: Client's CDM project arrangements/ checklist

CC [*The client*]

Planning supervisor's prequalification (CDM coordinator)

Project:

Name of organisation:

Address:

Contact for further information and tel. no.:

Email address:

With reference to the above project it is a statutory requirement that competence and adequacy of resources are established in compliance with regulations 8(1) and 9(1) of the Construction (Design and Management) Regulations 1994. In respect of this please complete, sign and provide all requisite information as requested below.

No.		Included Yes/No
1.	Please provide a signed copy of your organisation's health and safety policy (if you employ five or more).	
2.	Include details as to how you carry out all the duties of a planning supervisor. Provide details of how you embrace the CDM Approved Code of Practice.	
3.	Please ensure this includes specific arrangements for: a. ensuring health and safety management through all design phases; b. assisting the client and contractors with prequalification; c. developing a health and safety plan; d. assisting the client with the suitability of the principal contractor's health and safety plan; e. ensuring the health and safety file is undertaken.	
4.	Provide relevant experience and qualifications of the intended planning supervisor or planning supervisors, to include any training undertaken in the last three years.	
5.	Outline the procedures for monitoring, auditing and reviewing performance.	
6.	Provide a list of similar-type projects undertaken by the proposed planning supervisor.	
7.	Outline the proposed structure of the planning supervision service for this project.	
8.	What other health and safety resources does your organisation have in terms of planning supervision?	
9.	Has your organisation had any action taken against it by the Health and Safety Executive such as improvement notices, prohibition notices or successful prosecutions?	
10.	Please provide information relating to your professional indemnity insurance, e.g. a certificate.	

Name: Date:

Signature: Organisation:

Thank you for completing the prequalification. All information will be assessed by a competent person and feedback will be provided accordingly.

Designer's prequalification
(for CDM competence and resources)

Project:

Name of organisation:

Address:

Contact for further information and tel. no.:

Email address:

With reference to the above project it is a statutory requirement that competence and adequacy of resources are established in compliance with regulations 8(2) and 9(2) of the Construction (Design and Management) Regulations 1994. In respect of this please complete, sign and provide all requisite information as requested below.

No.		Included Yes/No
1.	Please provide a signed copy of your organisation's health and safety policy (if you employ five or more).	
2.	Include details as to how you carry out all the duties of a designer under the Construction (Design and Management) Regulations 1994. Provide details of how you embrace the CDM Approved Code of Practice.	
3.	Please ensure this includes specific arrangements for: a. ensuring the client is aware of his duties; b. designing with adequate regard to health and safety; c. providing relevant information to the project team; d. cooperating and communicating with the planning supervisor and other designers on the project.	
4.	Provide relevant experience and qualifications of the intended designer or designers, to include any relevant training undertaken in the last three years.	
5.	Outline the procedures for monitoring, auditing and reviewing performance.	
6.	Provide a list of similar-type projects undertaken.	
7.	Outline the proposed structure of the design service for this project.	
8.	What other health and safety resources does your organisation have in terms of health and safety management, for example reference books, guidance or sources of information?	
9.	Has your organisation had any action taken against it by the Health and Safety Executive such as improvement notices, prohibition notices or successful prosecutions?	
10.	Provide samples from previous projects of information passed to the planning supervisor during a project.	

Name: Date:

Signature: Organisation:

Thank you for completing the prequalification. All information will be assessed by a competent person and feedback will be provided accordingly.

Principal contractor's prequalification
(for CDM competence and resources)

Project:

Name of organisation:

Address:

Contact for further information and tel. no.:

Email address:

With reference to the above project it is a statutory requirement that competence and adequacy of resources are established in compliance with regulations 8(3) and 9(3) of the Construction (Design and Management) Regulations 1994. In respect of this please complete, sign and provide all requisite information as requested below.

No.		Included Yes/No
1.	Please provide a signed copy of your organisation's health and safety policy (if you employ five or more).	
2.	Include details as to how you carry out all the duties of a principal contractor under the Construction (Design and Management) Regulations. Provide details of how you embrace the CDM Approved Code of Practice.	
3.	Please ensure this includes specific arrangements for: a. developing the health and safety plan; b. ensuring designers and contractors engaged are both competent and adequately resourced; c. ensuring cooperation between contractors; d. providing information to the contractors and the planning supervisor; e. ensuring consultation with the workforce.	
4.	Provide relevant experience and qualifications of the intended team, to include any relevant training undertaken in the last three years.	
5.	Outline the procedures for monitoring, auditing and reviewing performance.	
6.	Provide a list of similar-type projects undertaken.	
7.	Outline the proposed health and safety management structure for this project.	
8.	What other health and safety resources does your organisation have in terms of health and safety management, for example safety advisers, reference books, guidance or sources of information?	
9.	Has your organisation had any action taken against it by the Health and Safety Executive such as improvement notices, prohibition notices or successful prosecutions?	
10.	Provide a sample construction phase health and safety plan from a previous project.	

Name: Date:

Signature: Organisation:

Thank you for completing the prequalification. All information will be assessed by a competent person and feedback will be provided accordingly.

Health and safety information survey

Note: Suitable for clients and planning supervisors developing a pre-tender/construction health and safety plan

Project...Date..............

Client details	
Site address	

Nature of construction work (brief description)

Timescale	Start:	Finish:

Existing drawings

Drawing no.	Scale	By	Description

Existing health and safety file: Yes/No	Location and access details

Existing environment

Subject	Details	Dwg ref.	Doc. ref.
Land use/other construction, etc.			
Services Gas: Water: Sewage: TV cable: Electricity: Telecoms: Fire alarm: Other:			
Traffic restrictions:			
Ground conditions:			
Construction materials:			
Site-wide elements: Schools:			

Shops: Other construction:			
Overlap With client's undertakings:			
Site rules:			
Continuing liaison:			

General project safety checklist

Existing hazards	Findings	Method statement required	Photo/ info required
Falls Roof access: Holes in floors: Balconies: High-level plant: Low windows: Low (below 910mm) ballast rails: Fixing points: Access equipment:			
Access (general) Car parking: Loading/ unloading: Security arrangements: Materials storage: Time restrictions:			
Structure stability Loose materials: Structural cracks: Supporting scaffold: Subsidence:			
Confined spaces Basements: Attic voids: Manhole access: Silos: Boiler rooms: Risers: Lift shafts: Ducting: Cooling tower:			
Plant and equipment Boiler: Cooking equipment: Air conditioning:			
Lifts:			

Asbestos ACMs	Yes/No	Date	Details
Has an asbestos survey been commissioned? Lab analyses: Air monitoring: Results: **Radon** Is premises situated in high-risk radon area? Air testing: Results:			
Legionellosis Surveyed: Cooling tower: Evaporative condensers: Spray humidifiers: HW system +300litres Showers: Fountains: Maintenance contract:			
Lead LCMs Lead paintwork: Roof flashing: Damp-proof courses: Has an assessment been undertaken?			
Damp/rot treatment Chemical injection: Chemicals sprayed: Paint-on chemicals: Dry rot spores:			
Electrical installation Current test certificate: Lift certificate:			
COSHH assessments Existing: Chemicals: Processes:			

Signed:		Date:	

Comments

Project Health and Safety File:

Certificate of handing over
CDM reg 14(f)

Project: ....

Client: ....

This certificate is signed in acknowledgement of the handing over of the health and safety file for the above project.

The health and safety file has been compiled in accordance with the requirements of the Construction (Design and Management) Regulations 1994 (as amended) and is completed within the terms of regulations 14(d) and 14(e) with the exception of the following information which has not been received by the planning supervisor:

Outstanding

....
due from..

....
due from..

....
due from..

This information will be added to the file by the client when received.

Signed............................. For .... (Planning supervisor)

Date of file
handover..

Signed............................. For.. (Client)

Please keep for record

Planning supervisor's questionnaire appraisal

Planning supervisor		Project	

No.	Questions	Assessment rating (tick ✓)						Sub-total
		0	**1**	**2**	**3**	**4**	**5**	
1	Health and safety policy							
2	Provision of qualifications							
3	Names of key personnel with related training							
4	Details of health and safety resources							
5	Details of health and safety training/seminars							
6	Coordination procedures for health and safety info							
7	Procedures for health and safety management in design							
8	Procedures for competence assessment							
9	Example of pre-construction health and safety plan							
10	Comments from HSE regarding PS service							
11	History of legal action							
12	Structure of anticipated PS team							
13	Info on scale of fees charged							

Action required:	0	1	2	3	4	5	Total
More information/ evidence: details	Unsatisfactory		Satisfactory pending action		Satisfactory		

....

....

....

....

....

....

....

....

....

Note: Planning supervisors MUST score over 40 to be considered competent and adequately resourced as required by the CDM Regulations. If a principal contractor does not score over 40, more information will be required or written evidence of action to be taken.

In my opinion the planning supervisor is competent and adequately resourced.

Signature of assessor responsible... Date...................................

Copy to:

Client information to planning supervisor checklist

Note: The CDM Regulations require that clients make available relevant health and safety information about existing structures and the site.

Information description	Location (if not available)	Tick where applicable✓
Existing site surveys		
Existing geology		
Structure surveys		
Construction history		
Contamination		
Site services		
Other buried services		
Overhead services		
Underground chambers, wells, water courses, etc.		
Existing building drawings		
Structure		
Previous structural alterations		
Fire damage		
Anchorage points		
Fragile materials		
Hazardous construction materials		
Use of pre-stressed or post-tensioned structures		
Existing building services surveys		
Asbestos survey/management plan		
Other hazardous materials		
Hazards from adjoining property		
Hazards from existing operations		
Existing O&M or building manuals		
Existing health and safety files		
Existing drawings		
Client rules, emergency procedures, requirements or restrictions		
Access restrictions		
Other persons on site: children/disabled/tenants/ shoppers/maintenance workers/delivery people/ other. Please specify below:		
Other (please specify)		

Copy to		Date:
Comments		

Note: Alternatively it may in some instances be beneficial to commission surveys to determine some of the detail in the list above, e.g. asbestos or geological surveys.

DESIGNER CDM CHECKLIST

Project name

Project ref. no.

Ref.	Project stage	CDM requirements	Considerations/guidance	Completed
1	Pre-contract	**Provide evidence of:** ▪ **competence; and** ▪ **adequate resources** As requested by the client	Based on the specific project you will be required to prove, to the client, that you are both competent and adequately resourced to effectively carry out your duties as per the CDM Regulations.	
2	On appointment	**Ensure that the client is aware of their duties under the CDM Regulations**	This is an opportunity to explain, especially to more non-professional construction clients, what is required of them. However, make no assumption that clients perceived to be more professional, such as government, housing associations or developers, have a complete understanding of their CDM responsibilities. *Refer to Form C1 (or provide Form C1 to the client)*	
3	Throughout all stages of design	**Ensure that, for your design, you have given adequate regard to the need to:** ▪ **avoid** ▪ **reduce** ▪ **control** ▪ **transfer** **all foreseeable risks**	You are required to consider the health and safety implications associated with constructing, maintaining and even demolishing or dismantling your design. This is based on your assessment of the associated hazards and the level of risk they pose. An appropriate risk assessment strategy will be necessary. Consider discussing an appropriate response with the planning supervisor.	

4	Throughout all stages of design	**Ensure the design has included adequate information**	This may be the output of your design risk assessment, notes on the drawing or indeed design methodology if necessary. Consider information that is required by competent contractors to build the structure safely, maintain the structure safely and even demolish or dismantle the building safely. Note that the planning supervisor will only be interested in significant residual risks for the health and safety plan and the health and safety file. You could discuss the provision of information with the project planning supervisor to establish an appropriate management arrangement.
5	Throughout all stages of design	**Cooperate with other designers on the project and cooperate with the planning supervisor**	Consider how you will interact with other designers on the project and discuss the communication/ working arrangements with the planning supervisor.
6	Design changes/ variations	**Consider the health and safety implications of any variations to the design and undertake a risk assessment**	Review existing design risk assessments and consider any changes. Inform the planning supervisor of any new significant residual risks.
7	During design	**Ensure that relevant information is provided to the planning supervisor for inclusion in the health and safety file**	Discuss with the planning supervisor the contributions you can make to the health and safety file. Examples include as built drawings, hazard data sheets, access issues, and replacement advice for awkward/unusual prefabricated elements.
8	Post-project	**Review the health and safety performance on the project and consider the contributions made with a view to continuous improvement and best practice**	Can good practices be shared across the organisation? Were any bad practices undertaken and if so why? How effective were the design team and planning supervisor in addressing health and safety?

continued overleaf

As designer/contract administrator I confirm that so far as is reasonably practicable all CDM designers' duties within my control have been complied with.

CDM sign-off		Date
Hazard	**Design action** (Please comment on how hazards have been successfully AVOIDED/REDUCED/CONTROLLED AND TRANSFERRED through good design)	
Informing client of CDM duties	**Action taken** (Please comment on how reasonable steps were taken to ensure the client was informed of their CDM duties. Note: This section should only be completed where difficulties or concerns were experienced regarding the client's appreciation of their duties and/or their competence)	
Summary of other problems	Lack of information from client or other designers to undertake effective risk assessment; problems obtaining suitable information for health and safety file.	
Copy to (for audit or policy development)		

Design risk assessment toolkit Project.........................

Design Stage.........................

Activity	Hazards	Population at risk	Risk			Design action, mitigation, information and construction/ maintenance risk control measures	Management of action		
			L	S	RF		Action by (name)	Sig./ res. risk	Check (sig.)

Assessment by			Date		Copy to *(please circle)* planning supervisor/external designer/engineer/ production manager/project sponsor/other (specify)

Ref. no. | **Key**: L = Likelihood (LMH–123) S = Severity (LMH–123) RF = Risk Factor (1–3 = Low; 4–6 = Medium; 7–9 = High)

PLANNING SUPERVISOR'S CDM CHECKLIST

Project name

Project ref. no.

Ref.	Project stage	CDM requirements	Considerations/guidance	Completed
1	Pre-design	Establish extent of service level agreement	The client and planning supervisor must clearly define who is to develop the health and safety file, whether advice is required on competence, resources and the suitability of the health and safety file, etc.	
2	Design	Ensure the F10 HSE notification is sent	No requirement to enter contractors' details. If not selected, HSE can be notified on additional notification.	
3	Design	Receive relevant information from client on the existing site hazards, activities and condition of the structure	Client should provide this information based on reasonable enquiries.	
4	All stages	Offer advice on designers' and contractors' competence and resources	Client to request service, which can be executed with questionnaires, interview or both. Report to client required.	
5	Design	Ensure designers design with adequate regard to health and safety and design to avoid, reduce, control and transfer hazards	Planning supervisor to ensure designers have a strategy to apply design hierarchy throughout design, and need evidence on project.	

6	Design	Ensure designers provide adequate health and safety-related design information in or with their designs	Designers to be advised of requirement not to provide information on issues and hazards that one would expect competent designers and contractors to be aware of, but only on unusual and difficult-to-manage issues.
7	Design	Ensure designers cooperate with other designers and planning supervisor	Evidence of design collaboration to address health and safety problems and provide action on risk assessment pro formas, design team minutes or risk registers.
8	Pre-tender	Produce pre-tender health and safety plan in accordance with CDM ACoP Appendix 3	Pre-tender plan to consist of designer and client information and restrictions.
9	Pre-tender	Report on suitability of construction phase health and safety plan	Must be suitably developed for work to start and may not contain arrangements for work programmed for later in the project. Should contain initial method statements.
10	All stages	Ensure development and handover of health and safety file	Development can start during design and will include checking collation of material against planned file structure.

continued overleaf

As project planning supervisor I confirm *that so far as is reasonably practicable all planning supervisor's duties within my control have been complied with.*

CDM sign-off		Date	
Hazard	**Design action** *(Please comment on how hazards have been successfully AVOIDED/REDUCED/CONTROLLED AND TRANSFERRED through good design)*		
CDM PS duties	**Best practice: action taken** *(Please comment on best practice options employed on this project to promote the efficient management of health and safety)*		
Summary of other problems	*Lack of information from client or other designers to undertake effective risk assessment; problems obtaining suitable information for health and safety file.*		
Copy to (for audit or policy development)			

Designer's questionnaire appraisal

Designer		Project	

No.	Questions	Assessment rating (tick ✓)						Sub-total
		0	**1**	**2**	**3**	**4**	**5**	
1	Health and safety policy							
2	Aware of CDM responsibilities.							
3	Names of/membership bodies, groups, etc.							
4	Names of key personnel with training							
5	Aware of duties under CDM reg 13(1)							
6	Design with adequate regard to health and safety							
7	Design risk assessment pro formas enclosed							
8	Aware of control hierarchy							
9	Procedures for assessment of designer competence							
10	Three examples of design risk assessments							
11	History of legal action							
12	Dissemination of health and safety info for update							
13	Organisation structure for project							
14	Designer access to CDM AcoP HSG224							
15	Coordination of health and safety info between designers							
16	Post-contract reviews							
17	QA for health and safety issues being signed off							
18	Proof of health and safety resources							
19	Specialist resources							
20	Health and safety library to support design health and safety							

Action required:	0	1	2	3	4	5	Total
More information/evidence: details	Unsatisfactory		Satisfactory Pending Action		Satisfactory		

More information/evidence: details
....
....
....
....
....
....
....
....
....

Note: *Designers MUST score over 70 to be considered competent and adequately resourced as required by the CDM Regulations. If a designer does not score over 70, more information will be required or written evidence of action to be taken.*

In my opinion the company/designer is competent and adequately resourced.

Signature of assessor responsible...Date......................

Copy to

Principal contractor's questionnaire appraisal

Principal contractor		Project	

No.	Questions	Assessment rating (tick ✓)						Sub-total
		0	1	2	3	4	5	
1	Health and safety policy							
2	Aware of CDM responsibilities							
3	Names of key personnel with related training							
4	History of legal action							
5	Details of health and safety awards							
6	Summary of accident history							
7	Details of health and safety adviser							
8	Examples of construction risk assessments							
9	Examples of construction phase health and safety plan							
10	Post-project reviews of health and safety management							
11	Organisation structure for coordination between parties							
12	Procedures for competence assessments							
13	Aware of PC need to provide health and safety info							
14	Consider organisation competent/ adequately resourced							
15	Aware construction phase plan to be developed							

Action required: More information/evidence: details	0	1	2	3	4	5	Total
	Unsatisfactory		Satisfactory pending action		Satisfactory		

....
....
....
....
....
....
....
....

Note: *Principal contractors MUST score over 55 to be considered competent and adequately resourced as required by the CDM Regulations. If a principal contractor does not score over 55, more information will be required or written evidence of action to be taken.*

In my opinion the principal contractor is competent and adequately resourced.

Signature of assessor responsible...Date..............................

Copy to

Construction phase plan assessment

Project:

Principal contractor:

Address:
Postcode

Contact for further information and tel. no.:

Email address:

Note: As you are aware under regulation 10 of the Construction (Design and Management) Regulations 1994 the client must ensure a construction phase health and safety plan is developed before work commences. The planning supervisor can offer advice to the client on this matter.

No.	Does the construction phase health and safety plan include:	Yes/No	Critical Yes/No
1	A project description that includes the activities.		
2	Contact details of CDM duty holders.		
3	Details of timescales including shutdown periods or different phases.		
4	Details of management structure and arrangements for health and safety.		
5	Named individuals and responsibilities for hazard/construction activities.		
6	Arrangements (where appropriate) for: • regular liaison between parties on site; • consultation with the workforce; • exchange of design information; • handling design changes; • selection and control of contractors; • exchange of health and safety information between contractors; • security, site induction and on-site training; • welfare facilities and first aid; • reporting and investigation of accidents; • production and approval of risk assessments and method statements.		
7	Clear site health and safety rules.		
8	Fire and emergency procedures.		
9	Arrangements (where appropriate) for controlling significant site risks for: • services, including temporary services/installations; • preventing falls; • working with or near fragile materials; • control of lifting operations; • maintenance of plant and equipment; • poor ground conditions; • traffic routes and segregation of vehicles and pedestrians; • storage of hazardous materials; • dealing with existing unstable structures; • accommodating adjacent land use; • entry into confined spaces; • dealing with/ removal of asbestos; • dealing with lead; • manual handling;		

	• use of hazardous substances; • dealing with contaminated land; • working over or near water; • reducing noise and vibration; • other(s) (specify)		
10	The health and safety file arrangements for: • layout and format; • collection and gathering of information; • storage of information.		

Is the plan suitably developed for work to start?   YES/NO

Summary of required development to deem the plan suitably developed for work to commence.

Signed

Date

Copy to: PC/client/agent/other...

Construction health and safety plan assessment

Certificate for work to commence
CDM regs 10 and 15(4)

I/We hereby confirm that .... as project planning supervisors have assessed the construction phase health and safety plan prepared by .... and dated .... for project ....

*The project was initially notified to the Health and Safety Executive under regulation 7(1) of the Construction (Design and Management) Regulations 1994 on ....
A copy of the notice is attached for signature.*

*I/We hereby advise the client that the plan has been developed sufficiently by the principal contractor in compliance with regulation 15(4) so that the **construction phase of the project may commence**.*

Signed..

For.. (planning supervisors)

Date..

Please keep for record

Project risk register

Form PS6

Project			Risk register coordinator			Tel. and email				

Item ref. no.	Date entered	Process/ location	Hazards	Action taken and date	Further action and date	By	Date	Completed

PRINCIPAL CONTRACTOR CDM CHECKLIST

Project name

Project ref. no.

Ref.	Project stage	CDM requirements	Considerations/guidance	Completed
1	Pre-construction	Sign and provide information for additional F10	Information on contractors' details to be supplied to the planning supervisor (or person responsible) for inclusion in F10 if known.	
2	Pre-construction	Develop construction phase health and safety plan and provide to client or planning supervisor for assessment	Plan only has to be suitably developed and should contain initial method statements. Only start work when approved.	
3	Construction	Ensure cooperation between contractors to facilitate effective health and safety planning and execution on site	By setting up coordination meetings and arrangements to share information on health and safety issues associated with their construction activities.	
4	Construction	Ensure every contractor complies with the site rules and conditions of the construction phase health and safety plan	Monitoring arrangements required to check compliance and report findings and remedial actions.	
5	Construction	Ensure only authorised persons are allowed on site	Security and access control arrangements required, e.g. signing-in/-out procedure, induction/authorisation stickers on hard hats and security cards.	
6	Construction	Display copy of F10 HSE notification on site	F10 to be displayed in a prominent place on site.	

continued overleaf

7	Construction	Provide planning supervisor with relevant health and safety information for the health and safety file	File information to be checked off against the file's agreed structure.
8	Construction	Give reasonable instructions to contractors	As an output of coordination meetings contractors must be controlled to implement the construction programme and ensure health and safety by avoiding interfaces on site that increase risk and cause accidents and ill health.
9	Construction	Bring the site rules to the attention of all persons on site.	Induction and health and safety awareness training and safety poster are methods of communicating site rules to people on site.
10	Construction	Provide every contractor with relevant information and training on health and safety	Risk assessments and method statements provide project-specific information, but design risk assessments, surveys and safety reports also help inform the contractor. Training over and above induction can include awareness, toolbox talks and skills training, e.g. CITB/CPCS for plant operatives.
11	Construction	Ensure employees and the self-employed can discuss health and safety and offer advice and also that information is coordinated between all on site	Structured approach to this is safety forum meetings, open-discussion toolbox talks and a promotion at induction to encourage personnel to discuss health and safety and report hazards or hazardous situations.

As project construction manager I confirm that so far as is reasonably practicable all CDM principal contractor's duties within my control have been complied with.

CDM sign-off		Date
Hazard	**Site management action** (*Please comment on how hazards have been successfully AVOIDED/ REDUCED/ CONTROLLED through good health and safety management.* **Note:** *This section only applies to unusual and difficult-to-manage hazards. This could include deep excavations, installing awkward components, working with hazardous substances including asbestos, difficult traffic management issues, high work, working in confined spaces, hot work, etc.*)	
Controlling and informing contractors	**Action taken** (*Please comment on how reasonable steps were taken to ensure the contractor was informed of the site rules, unusual and difficult-to-manage hazards, safety arrangements on site, etc.* **Note:** *This section should only be completed where unusual, high-risk and difficult-to-manage hazards are present*)	
Summary of other problems	*Lack of information from the planning supervisor or other designers to undertake effective risk assessment and manage site safety; problems obtaining suitable information for health and safety file. List what information you would appreciate in the pre-construction health and safety plan.*	
Copy to safety adviser (for audit or policy development)	Sent to	Date sent

continued overleaf

Contractor's prequalification
(for CDM competence and resources)

Project:
Name of organisation:
Address:
Contact for further information and tel. no.:
Email address:

With reference to the above project it is a statutory requirement that competence and adequacy of resources are established in compliance with regulations 8(3) and 9(3) of the Construction (Design and Management) Regulations 1994. In respect of this please complete, sign and provide all requisite information as requested below.

No.		Included Yes/No
1.	Please provide a signed copy of your organisation's health and safety policy (if you employ five or more).	
2.	Include details as to how you carry out all the duties of a contractor under the Construction (Design and Management) Regulations. Provide details of how you embrace the CDM Approved Code of Practice.	
3.	Please ensure this includes specific arrangements for: a. ensuring designers and contractors engaged are both competent and adequately resourced; b. cooperating with the principal contractor; c. providing information to the principal contractor on risks associated with their undertakings; d. complying with relevant rules from the principal contractor and the health and safety plan.	
4.	Provide relevant experience and qualifications of the intended team, to include any relevant training undertaken in the last three years.	
5.	Outline the procedures for monitoring, auditing and reviewing performance.	
6.	Provide a list of similar-type projects or work undertaken.	
7.	Outline the proposed health and safety management approach, for example risk assessments, method statements and lifting plans. Please provide samples.	
8.	What other health and safety resources does your organisation have in terms of health and safety management, for example safety advisers, reference books, guidance or sources of information?	
9.	Has your organisation had any action taken against it by the Health and Safety Executive such as improvement notices, prohibition notices or successful prosecutions?	
10.	Please provide accident/incident statistics for the last three years.	

Name: Date:

Signature: Organisation:

Thank you for completing the prequalification. All information will be assessed by a competent person and feedback will be provided accordingly.

Contractor's questionnaire feedback appraisal

Contractor		Project	

No.	Questions	Assessment rating (tick ✓)						Sub-total
		0	1	2	3	4	5	
1	Health and safety policy document							
2	Health and safety management procedures							
3	Professional safety adviser							
4	Project safety manager							
5	Examples of risk assessments							
6	Examples of method statements							
7	History of enforcement action							
8	History of accidents							
9	Monitoring of health and safety standards							
10	Details of health and safety training							

	0	1	2	3	4	5	Total
	Unsatisfactory		Satisfactory pending action		Satisfactory		

Action required:

More information/evidence: details

....

....

....

....

Need to attend training: details

....

....

....

....

....

Note: *Contractors MUST score over 35 to be considered eligible for induction training and access to site. If a contractor does not score over 35, more information will be required or written evidence of action to be taken before they start work.*

In my opinion the company/contractor is competent and adequately resourced to start work following site induction training.

Signature of assessor.. Date.................................

Signature of project manager... Date.................................

I/My company will attend site induction training and follow the site rules and any instructions given by the principal contractor on site.

Signature of contractor.. Date.................................

Copy to

Construction phase plan contents checklist

Project	Date

Checked by (principal contractor representative)

Note: *Please check the construction phase plan against the checklist below before submitting it to the client or planning supervisor for approval. Please amend the plan if necessary.*

No.	Does the construction phase health and safety plan include:	Yes/No	Plan changed
1	A project description that includes the activities.		
2	Contact details of CDM duty holders.		
3	Details of timescales including shutdown periods or different phases.		
4	Details of management structure and arrangements for health and safety.		
5	Named individuals and responsibilities for hazard/construction activities.		
6	Arrangements (where appropriate) for: • regular liaison between parties on site; • consultation with the workforce; • exchange of design information; • handling design changes; • selection and control of contractors; • exchange of health and safety information between contractors; • security, site induction and on-site training; • welfare facilities and first aid; • reporting and investigation of accidents; • production and approval of risk assessments and method statements.		
7	Clear site health and safety rules.		
8	Fire and emergency procedures.		
9	Arrangements (where appropriate) for controlling significant site risks for: • services, including temporary services/installations; • preventing falls; • working with or near fragile materials; • control of lifting operations; • maintenance of plant and equipment; • poor ground conditions; • traffic routes and segregation of vehicles and pedestrians; • storage of hazardous materials; • dealing with existing unstable structures; • accommodating adjacent land use; • entry into confined spaces; • dealing with/removal of asbestos; • dealing with lead; • manual handling; • use of hazardous substances; • dealing with contaminated land; • working over or near water; • reducing noise and vibration; • other(s) specify..		
10	The health and safety file arrangements for: • layout and format; • collection and gathering of information; • storage of information.		

Is the plan suitably developed for submission to the client/planning supervisor YES/NO	
Signed	Date
Note: Copy on project file	

Health and safety risk assessment and method appraisal

Project:		Submission date:	
Contractor:		RA ref no.	MS ref. no.
Method statement title:			

The risk assessments and/or method statements have been assessed by:

.... and the following status allocated.

Please enter status A, B or C clearly in the box below:

A = Works may proceed

B = Works may proceed subject to comments below
and submission of revised method statement/ risk
assessment(s) within 24 hours

C = **Rejection** Do not proceed with the works – contact construction manager

Comments	Action

I confirm that I have assessed the suitability of the risk assessments and/or method statements above: Sign here .... Site manager

Site manager/agent responsible: print name		Date:	
Copy to contract manager and subcontractor Yes/No			

Site safety coordination register

Project

Ref no.

Contracts manager

Construction activity and reference no.	Contractor(s)	Start date	Pre-qualification questionnaire		Initial induction training given	Contractors briefed on specific activities and arrangements	Risk assessments/ method statement approval (initials)			Safety information given by PC to contractors			Authorised to start (sign)
			Sent	Received and accepted			By	Ref no.	Appro ved	Risk assess ment	Drawings	Method statement	

CDM audit pack

Form AP1

Key:
✓ = suitably addressed
Ref no. = details below
NA = not applicable on this project

Project/ audit date
Title and ref no.

A project: ref. no. (456/ss21)

Group	Duty item	A (456/ss21)									
Client's duties (*3 & *5 can be following advice from PS)	Allowance of adequate time *1	✓									
	Appointment of des., PS and PC *2	✓									
	All parties competent *3	✓									
	Provision H&S information *4	✓									
	Suitability cons. phase plan *5	✓									
	Availability of H&S file *6	✓									
Designer's duties	Informed client of CDM duties *7	✓									
	Design risk assessment *8	✓									
	Additional information *9	✓									
	Cooperation with PS and des. *10	✓									
Planning supervisor's duties	F10 HSE notification *11	✓									
	Pre-construction phase plan *12	✓									
	Ensuring designers follow ERIC (Elimination, Reduction, Information, Control) *13	A21									
	Ensuring designers give info *14	✓									
	Ensuring designers cooperate *15	✓									
	Addressed *3 and *5 for client *16	✓									
	H&S file development *17	✓									
Principal contractor's duties	Construction phase plan *18	✓									
	File info to PS *19	✓									
	PC address reg. 13 (designer) *20	✓									
	Directions to contractors *21	✓									

(Comments Stage 2)

continued overleaf

Project	Ref. no.	Comments/CDM weaknesses/good practice	Action	By	Date
A project	A21	Poor compliance with CDM reg. 13 in that a suitable and sufficient risk assessment was not undertaken resulting in an element of design that was difficult to construct/access for maintenance, etc.	In-house training for designer using company training PowerPoint and test procedure. Designer to be monitored by lead designer. Amendments to TCDM database to include this element of design/advise designers. Email to design team highlighting issues and actions.		
		Good practice identified in sharing health and safety info with other external designer that was discussed in review meeting resulting in XYZ.	TCDM database upgraded to include XYZ and email to all designers and green list updated to include best practice.		

Project audit guidance notes

Action point *	CDM policy section and CDM ACoP	Guidance for auditor
*1	Allowance of adequate time (CDM ACoP)	ACoP requires client to allow adequate time for design, planning, preparation and construction work. Audit may require evaluating records of meetings and interviewing parties.
*2	Appointment of des., PS and PC (CDM reg. 6)	Look for formal appointment with objective that all project parties are aware of who is responsible for what. Audit may require evaluating records of meetings and interviewing parties.
*3	All parties competent (CDM regs 8 and 9)	Will need to see both competence and resource questionnaire and appraisal of feedback or record of meeting to assess competence. Duty of client, but generally arrangements authorise planning supervisor.
*4	Provision of H&S information (CDM reg. 11)	Evidence of appropriate information given to the planning supervisor in respect of the structure, hazards, environment, etc.
*5	Suitability cons. phase plan (CDM reg. 10)	Suitability based on an appropriate response in relation to the significant risks. Plan should also have arrangements for the plan's management, implementation and development. Duty of client, but generally arrangements authorise planning supervisor.
*6	Availability of H&S file (CDM reg. 12)	Assessment based on the existing file information being available to designers as necessary and evidence of client's arrangements for availability.
*7	Informed client of CDM duties (CDM reg. 13(1))	Where client is inexperienced and obviously not aware of how to address CDM, evidence of approach is necessary.
*8	Design risk assessment and best practice (CDM reg. 13(2))	Evidence of hazard identification, application of control hierarchy (avoidance, reduction and control) and good design practice required.
*9	Additional information (CDM reg. 13(2))	Evidence of appropriate information given in respect of supporting the

		health and safety design issues required. Appropriate information based also on CDM ACoP, para 131.
*10	Cooperation with PS and des. (CDM reg. 13(2))	Measure suitability against pre-construction plan and records of design review meetings. Interviews with parties can also provide necessary information. May need to check risk assessments/risk register.
*11	F10 HSE notification (CDM Reg 7)	Check document (F10) and contents for accuracy.
*12	Pre-construction phase plan (CDM reg. 15)	Measure suitability against client information and design risk assessment, and also Appendix 3 of CDM ACoP and nature and risk profile of project.
*13	Ensuring designers ERIC (CDM reg. 14)	Evaluation of design risk assessments or interviews with lead designers.
*14	Ensuring designers give info (CDM reg. 14)	Measure suitability against pre-construction plan.
*15	Ensuring designers cooperate (CDM reg. 14)	Measure suitability against pre-construction plan and records of design review meetings. Interviews with parties can also provide necessary information.
*16	Addressed *3 and *5 for client (CDM regs 8, 9 and 10)	Assessment of competence: will need to see both competence and resource questionnaire and appraisal of feedback or record of meeting to assess competence. Suitability of construction phase plan: suitability based on an appropriate response in relation to the significant risks. Plan should also have arrangements for the plan's management, implementation and development.
*17	H&S file development (CDM reg. 14)	Appraisal of project health and safety file. Measure information supplied against arrangements for file in safety plans and evidence in the form of the planning supervisor's input and project communications.
*18	Construction phase plan (CDM reg. 15)	Measure suitability against pre-construction plan, Appendix 3 of CDM ACoP and nature and risk profile of project.

*19	File info to PS (CDM reg. 16)	Appraisal of project health and safety file. Measure information supplied against arrangements for file in safety plans and evidence in the form of the planning supervisor's input and project communications.
*20	PC address reg. 13 (designer)	May obtain info from construction phase plan, interviews, risk assessments, records of design review meetings, etc.
*21	Directions and info to contractors (CDM regs 16 and 17)	May obtain info from construction phase plan, interviews, etc.

The Construction (Design and Management) Regulations 1994

Whereas the Health and Safety Commission has submitted to the Secretary of State, under section 11(2)(d) of the Health and Safety at Work etc. Act 1974 ('the 1974 Act'), proposals for the purpose of making regulations after the carrying out by the said Commission of consultations in accordance with section 50(3) of the 1974 Act;

And whereas the Secretary of State has made modifications to the said proposals under section 50(1) of the 1974 Act and has consulted the said Commission thereon in accordance with section 50(2) of that Act;

Now therefore, the Secretary of State, in exercise of the powers conferred on him by sections 15(1), (2), (3)(a) and (c), (4)(a), (6)(b) and (9), and 82(3)(a) of, and paragraphs 1(1)(c), 6(1), 14, 15(1), 20 and 21 of Schedule 3 to, the 1974 Act, and of all other powers enabling him in that behalf and for the purpose of giving effect to the said proposals of the said Commission with modifications as aforesaid, hereby makes the following Regulations:

Citation and commencement

1. These Regulations may be cited as the Construction (Design and Management) Regulations 1994 and shall come into force on 31st March 1995.

Interpretation

2.—(1) In these Regulations, unless the context otherwise requires –

'agent' in relation to any client means any person who acts as agent for a client in connection with the carrying on by the person of a trade, business or other undertaking (whether for profit or not);

'cleaning work' means the cleaning of any window or any transparent or translucent wall, ceiling or roof in or on a structure where such cleaning involves a risk of a person falling more than 2 metres;

'client' means any person for whom a project is carried out, whether it is carried out by another person or is carried out in-house;

'construction phase' means the period of time starting when construction work in any project starts and ending when construction work in that project is completed;

'construction work' means the carrying out of any building, civil engineering or engineering construction work and includes any of the following –

(a) the construction, alteration, conversion, fitting out, commissioning, renovation, repair, upkeep, redecoration or other maintenance (including cleaning which involves the use of water or an abrasive at high pressure or the use of substances classified as corrosive or toxic for the purposes of regulation 7 of the Chemicals (Hazard Information and Packaging) Regulations 1993), de-commissioning, demolition or dismantling of a structure,

(b) the preparation for an intended structure, including site clearance, exploration, investigation (but not site survey) and excavation, and laying or installing the foundations of the structure,

(c) the assembly of prefabricated elements to form a structure or the disassembly of prefabricated elements which, immediately before such disassembly, formed a structure,

(d) the removal of a structure or part of a structure or of any product or waste resulting from demolition or dismantling of a structure or from disassembly of prefabricated elements which, immediately before such disassembly, formed a structure, and

(e) the installation, commissioning, maintenance, repair or removal of mechanical, electrical, gas, compressed air, hydraulic, telecommunications, computer or similar services which are normally fixed within or to a structure,

but does not include the exploration for or extraction of mineral resources or activities preparatory thereto carried out at a place where such exploration or extraction is carried out;

'contractor' means any person who carries on a trade, business or other undertaking (whether for profit or not) in connection with which he –

(a) undertakes to or does carry out or manage construction work,

(b) arranges for any person at work under his control (including, where he is an employer, any employee of his) to carry out or manage construction work;

'design' in relation to any structure includes drawing, design details, specification and bill of quantities (including specification of articles or substances) in relation to the structure;

'designer' means any person who carries on a trade, business or other undertaking in connection with which he –
(a)   prepares a design, or
(b)   arranges for any person under his control (including, where he is an employer, any employee of his) to prepare a design,
relating to a structure or part of a structure;

'developer' shall be construed in accordance with regulation 5(1);

'domestic client' means a client for whom a project is carried out not being a project carried out in connection with the carrying on by the client of a trade, business or other undertaking (whether for profit or not);

'health and safety file' means a file, or other record in permanent form, containing the information required by virtue of regulation 14(d);

'health and safety plan' means the plan prepared by virtue of regulation 15;

'planning supervisor' means any person for the time being appointed under regulation 6(1)(a);

'principal contractor' means any person for the time being appointed under regulation 6(1)(b);

'project' means a project which includes or is intended to include construction work;

'structure' means –
(a)   any building, steel or reinforced concrete structure (not being a building), railway line or siding, tramway line, dock, harbour, inland navigation, tunnel, shaft, bridge, viaduct, waterworks, reservoir, pipe or pipe-line (whatever, in either case, it contains or is intended to contain), cable, aqueduct, sewer, sewage works, gasholder, road, airfield, sea defence works, river works, drainage works, earthworks, lagoon, dam, wall, caisson, mast, tower, pylon, underground tank, earth retaining structure, or structure designed to preserve or alter any natural feature, and any other structure similar to the foregoing, or
(b)   any formwork, falsework, scaffold or other structure designed or used to provide support or means of access during construction work, or

(c) any fixed plant in respect of work which is installation, commissioning, de-commissioning or dismantling and where any such work involves a risk of a person falling more than 2 metres.

(2) In determining whether any person arranges for a person (in this paragraph called 'the relevant person') to prepare a design or to carry out or manage construction work regard shall be had to the following, namely –

(a) a person does arrange for the relevant person to do a thing where –
   (i) he specifies in or in connection with any arrangement with a third person that the relevant person shall do that thing (whether by nominating the relevant person as a subcontractor to the third person or otherwise), or
   (ii) being an employer, it is done by any of his employees in-house;

(b) a person does not arrange for the relevant person to do a thing where –
   (i) being a self-employed person, he does it himself or, being in partnership it is done by any of his partners; or
   (ii) being an employer, it is done by any of his employees otherwise than in-house, or
   (iii) being a firm carrying on its business anywhere in Great Britain whose principal place of business is in Scotland, it is done by any partner in the firm; or
   (iv) having arranged for a third person to do the thing, he does not object to the third person arranging for it to be done by the relevant person,

and the expressions 'arrange' and 'arranges' shall be construed accordingly.

(3) For the purposes of these Regulations –

(a) a project is carried out in-house where an employer arranges for the project to be carried out by an employee of his who acts, or by a group of employees who act, in either case, in relation to such a project as a separate part of the undertaking of the employer distinct from the part for which the project is carried out; and

(b) construction work is carried out or managed in-house where an employer arranges for the construction work to be carried out or managed by an employee of his who acts or by a group of employees who act, in either case, in relation to such construction work as a separate part of the undertaking of the employer distinct from the part for which the construction work is carried out or managed; and

(c) a design is prepared in-house where an employer arranges for the design to be prepared by an employee of his who acts, or by a group of employees who act, in either case, in relation to such design as a separate part of the undertaking of the employer distinct from the part for which the design is prepared.

(4) For the purposes of these Regulations, a project is notifiable if the construction phase –

(a) will be longer than 30 days; or
(b) will involve more than 500 person days of construction work, and the expression 'notifiable' shall be construed accordingly.

(5) Any reference in these Regulations to a person being reasonably satisfied –

(a) as to another person's competence is a reference to that person being satisfied after the taking of such steps as it is reasonable for that person to take (including making reasonable enquiries or seeking advice where necessary) to satisfy himself as to such competence; and
(b) as to whether another person has allocated or will allocate adequate resources is a reference to that person being satisfied that after the taking of such steps as it is reasonable for that person to take (including making reasonable enquiries or seeking advice where necessary) –

(i) to ascertain what resources have been or are intended to be so allocated; and
(ii) to establish whether the resources so allocated or intended to be allocated are adequate.

(6) Any reference in these Regulations to –

(a) a numbered regulation or Schedule is a reference to the regulation in or Schedule to these Regulations so numbered; and
(b) a numbered paragraph is a reference to the paragraph so numbered in the regulation in which the reference appears.

Application of regulations

3.—(1) Subject to the following paragraphs of this regulation, these Regulations shall apply to and in relation to construction work.

(2) Subject to paragraph (3), regulations 4 to 12 and 14 to 19 shall not apply to or in relation to construction work included in a project where the client has reasonable grounds for believing that –

(a) the project is not notifiable; and
(b) the largest number of persons at work at any one time carrying out construction work included in the project will be or, as the case may be, is less than 5.

(3) These Regulations shall apply to and in relation to construction work which is the demolition or dismantling of a structure notwithstanding paragraph (2).

(4) These Regulations shall not apply to or in relation to construction work in respect of which the local authority within the meaning of regulation 2(1)

of the Health and Safety (Enforcing Authority) Regulations 1989 is the enforcing authority.

(5) Regulation 14(b) shall not apply to projects in which no more than one designer is involved.

(6) Regulation 16(1)(a) shall not apply to projects in which no more than one contractor is involved.

(7) Where construction work is carried out or managed in-house or a design is prepared in-house, then, for the purposes of paragraphs (5) and (6), each part of the undertaking of the employer shall be treated as a person and shall be counted as a designer or, as the case may be, contractor, accordingly.

(8) Except where regulation 5 applies, regulations 4, 6, 8 to 12 and 14 to 19 shall not apply to or in relation to construction work included or intended to be included in a project carried out for a domestic client.

Clients and agents of clients

4.—(1) A client may appoint an agent or another client to act as the only client in respect of a project and where such an appointment is made the provisions of paragraphs (2) to (5) shall apply.

(2) No client shall appoint any person as his agent under paragraph (1) unless the client is reasonably satisfied that the person he intends to appoint as his agent has the competence to perform the duties imposed on a client by these Regulations.

(3) Where the person appointed under paragraph (1) makes a declaration in accordance with paragraph (4), then, from the date of receipt of the declaration by the Executive, such requirements and prohibitions as are imposed by these Regulations upon a client shall apply to the person so appointed (so long as he remains as such) as if he were the only client in respect of that project.

(4) A declaration in accordance with this paragraph –

(a) is a declaration in writing, signed by or on behalf of the person referred to in paragraph (3), to the effect that the client or agent who makes it will act as client for the purposes of these Regulations; and

(b) shall include the name of the person by or on behalf of whom it is made, the address where documents may be served on that person and the address of the construction site; and

(c) shall be sent to the Executive.

(5) Where the Executive receives a declaration in accordance with paragraph (4), it shall give notice to the person by or on behalf of whom the declaration is made and the notice shall include the date the declaration was received by the Executive.

(6) Where the person referred to in paragraph (3) does not make a declaration

in accordance with paragraph (4), any requirement or prohibition imposed by these Regulations on a client shall also be imposed on him but only to the extent it relates to any matter within his authority.

Requirements on developer

5.—(1) This regulation applies where the project is carried out for a domestic client and the client enters into an arrangement with a person (in this regulation called 'the developer') who carries on a trade, business or other undertaking (whether for profit or not) in connection with which –

(a) land or an interest in land is granted or transferred to the client; and
(b) the developer undertakes that construction work will be carried out on the land; and
(c) following the construction work, the land will include premises which, as intended by the client, will be occupied as a residence.

(2) Where this regulation applies, with effect from the time the client enters into the arrangement referred to in paragraph (1), the requirements of regulations 6 and 8 to 12 shall apply to the developer as if he were the client.

Appointments of planning supervisor and principal contractor

6.—(1) Subject to paragraph (6)(b), every client shall appoint –

(a) a planning supervisor; and
(b) a principal contractor,
in respect of each project.

(2) The client shall not appoint as principal contractor any person who is not a contractor.
(3) The planning supervisor shall be appointed as soon as is practicable after the client has such information about the project and the construction work involved in it as will enable him to comply with the requirements imposed on him by regulations 8(1) and 9(1).
(4) The principal contractor shall be appointed as soon as is practicable after the client has such information about the project and the construction work involved in it as will enable the client to comply with the requirements imposed on him by regulations 8(3) and 9(3) when making an arrangement with a contractor to manage construction work where such arrangement consists of the appointment of the principal contractor.
(5) The appointments mentioned in paragraph (1) shall be terminated, changed or renewed as necessary to ensure that those appointments remain filled at all times until the end of the construction phase.

(6) Paragraph (1) does not prevent –

(a) the appointment of the same person as planning supervisor and as principal contractor provided that person is competent to carry out the functions under these Regulations of both appointments; or

(b) the appointment of the client as planning supervisor or as principal contractor or as both, provided the client is competent to perform the relevant functions under these Regulations.

Notification of project

7.—(1) The planning supervisor shall ensure that notice of the project in respect of which he is appointed is given to the Executive in accordance with paragraphs (2) to (4) unless the planning supervisor has reasonable grounds for believing that the project is not notifiable.

(2) Any notice required by paragraph (1) shall be given in writing or in such other manner as the Executive may from time to time approve in writing and shall contain the particulars specified in paragraph (3) or, where applicable, paragraph (4) and shall be given at the times specified in those paragraphs.

(3) Notice containing such of the particulars specified in Schedule 1 as are known or can reasonably be ascertained shall be given as soon as is practicable after the appointment of the planning supervisor.

(4) Where any particulars specified in Schedule 1 have not been notified under paragraph (3), notice of such particulars shall be given as soon as is practicable after the appointment of the principal contractor and, in any event, before the start of construction work.

(5) Where a project is carried out for a domestic client then, except where regulation 5 applies, every contractor shall ensure that notice of the project is given to the Executive in accordance with paragraph (6) unless the contractor has reasonable grounds for believing that the project is not notifiable.

(6) Any notice required by paragraph (5) shall –

(a) be in writing or such other manner as the Executive may from time to time approve in writing;

(b) contain such of the particulars specified in Schedule 1 as are relevant to the project; and

(c) be given before the contractor or any person at work under his control starts to carry out construction work.

Competence of planning supervisor, designers and contractors

8.—(1) No client shall appoint any person as planning supervisor in respect of a project unless the client is reasonably satisfied that the person he intends to appoint has the competence to perform the functions of planning supervisor under these Regulations in respect of that project.

(2) No person shall arrange for a designer to prepare a design unless he is reasonably satisfied that the designer has the competence to prepare that design.

(3) No person shall arrange for a contractor to carry out or manage construction work unless he is reasonably satisfied that the contractor has the competence to carry out or, as the case may be, manage, that construction work.

(4) Any reference in this regulation to a person having competence shall extend only to his competence –

(a) to perform any requirement; and

(b) to conduct his undertaking without contravening any prohibition,

imposed on him by or under any of the relevant statutory provisions.

Provision for health and safety

9.—(1) No client shall appoint any person as planning supervisor in respect of a project unless the client is reasonably satisfied that the person he intends to appoint has allocated or, as appropriate, will allocate adequate resources to enable him to perform the functions of planning supervisor under these Regulations in respect of that project.

(2) No person shall arrange for a designer to prepare a design unless he is reasonably satisfied that the designer has allocated or, as appropriate, will allocate adequate resources to enable the designer to comply with regulation 13.

(3) No person shall arrange for a contractor to carry out or manage construction work unless he is reasonably satisfied that the contractor has allocated or, as appropriate, will allocate adequate resources to enable the contractor to comply with the requirements and prohibitions imposed on him by or under the relevant statutory provisions.

Start of construction phase

10. Every client shall ensure, so far as is reasonably practicable, that the construction phase of any project does not start unless a health and safety plan complying with regulation 15(4) has been prepared in respect of that project.

Client to ensure information is available

11.—(1) Every client shall ensure that the planning supervisor for any project carried out for the client is provided (as soon as is reasonably practicable but in any event before the commencement of the work to which the information relates) with all information mentioned in paragraph (2) about the state or condition of any premises at or on which construction work included or intended to be included in the project is or is intended to be carried out.

(2) The information required to be provided by paragraph (1) is information which is relevant to the functions of the planning supervisor under these Regulations and which the client has or could ascertain by making enquiries which it is reasonable for a person in his position to make.

Client to ensure health and safety file is available for inspection

12.—(1) Every client shall take such steps as it is reasonable for a person in his position to take to ensure that the information in any health and safety file which has been delivered to him is kept available for inspection by any person who may need information in the file for the purpose of complying with the requirements and prohibitions imposed on him by or under the relevant statutory provisions.

(2) It shall be sufficient compliance with paragraph (1) by a client who disposes of his entire interest in the property of the structure if he delivers the health and safety file for the structure to the person who acquires his interest in the property of the structure and ensures such person is aware of the nature and purpose of the health and safety file.

Requirements on designer

13.—(1) Except where a design is prepared in-house, no employer shall cause or permit any employee of his to prepare, and no self-employed person shall prepare, a design in respect of any project unless he has taken reasonable steps to ensure that the client for that project is aware of the duties to which the client is subject by virtue of these Regulations and of any practical guidance issued from time to time by the Commission with respect to the requirements of these Regulations.

(2) Every designer shall –

    (a) ensure that any design he prepares and which he is aware will be used for the purposes of construction work includes among the design considerations adequate regard to the need –

      (i) to avoid foreseeable risks to the health and safety of any person at work carrying out construction work or cleaning work in or on the structure at any time, or of any person who may be affected by the work of such a person at work,

      (ii) to combat at source risks to the health and safety of any person at work carrying out construction work or cleaning work in or on the structure at any time, or of any person who may be affected by the work of such a person at work, and

      (iii) to give priority to measures which will protect all persons at work who may carry out construction work or cleaning work at any time and all persons who may be affected by the work of such persons at work over measures which only protect each person carrying out such work;

(b) ensure that the design includes adequate information about any aspect of the project or structure or materials (including articles or substances) which might affect the health or safety of any person at work carrying out construction work or cleaning work in or on the structure at any time or of any person who may be affected by the work of such a person at work; and

(c) co-operate with the planning supervisor and with any other designer who is preparing any design in connection with the same project or structure so far as is necessary to enable each of them to comply with the requirements and prohibitions placed on him in relation to the project by or under the relevant statutory provisions.

(3) Sub-paragraphs (a) and (b) of paragraph (2) shall require the design to include only the matters referred to therein to the extent that it is reasonable to expect the designer to address them at the time the design is prepared and to the extent that it is otherwise reasonably practicable to do so.

Requirements on planning supervisor

14. The planning supervisor appointed for any project shall –

(a) ensure, so far as is reasonably practicable, that the design of any structure comprised in the project –

    (i) includes among the design considerations adequate regard to the needs specified in heads (i) to iii) of regulation 13(2)(a), and

    (ii) includes adequate information as specified in regulation 13(2)(b);

(b) take such steps as it is reasonable for a person in his position to take to ensure co-operation between designers so far as is necessary to enable each designer to comply with the requirements placed on him by regulation 13;

(c) be in a position to give adequate advice to –

    (i)   any client and any contractor with a view to enabling each of them to comply with regulations 8(2) and 9(2), and to

    (ii) any client with a view to enabling him to comply with regulations 8(3), 9(3) and 10;

(d) ensure that a health and safety file is prepared in respect of each structure comprised in the project containing –

    (i)   information included with the design by virtue of regulation 13(2)(b), and

    (ii) any other information relating to the project which it is reasonably foreseeable will be necessary to ensure the health and safety of any person at work who is carrying out or will carry out construction work or cleaning work in or on the structure or of any person who may be affected by the work of such a person at work;

(e) review, amend or add to the health and safety file prepared by virtue of sub-paragraph (d) of this regulation as necessary to ensure that it contains the information mentioned in that sub-paragraph when it is delivered to the client in accordance with sub-paragraph (f) of this regulation; and

(f) ensure that, on the completion of construction work on each structure comprised in the project, the health and safety file in respect of that structure is delivered to the client.

Requirements relating to the health and safety plan

15.—(1) The planning supervisor appointed for any project shall ensure that a health and safety plan in respect of the project has been prepared no later than the time specified in paragraph (2) and contains the information specified in paragraph (3).

(2) The time when the health and safety plan is required by paragraph (1) to be prepared is such time as will enable the health and safety plan to be provided to any contractor before arrangements are made for the contractor to carry out or manage construction work.

(3) The information required by paragraph (1) to be contained in the health and safety plan is –

    (a) a general description of the construction work comprised in the project;

    (b) details of the time within which it is intended that the project, and any intermediate stages, will be completed;

    (c) details of risks to the health or safety of any person carrying out the construction work so far as such risks are known to the planning supervisor or are reasonably foreseeable;

    (d) any other information which the planning supervisor knows or could

ascertain by making reasonable enquiries and which it would be necessary for any contractor to have if he wished to show –

(i) that he has the competence on which any person is required to be
reasonably satisfied by regulation 8, or

(ii) that he has allocated or, as appropriate, will allocate, adequate
resources on which any person is required to be reasonably satisfied by regulation 9;

(e) such information as the planning supervisor knows or could ascertain
by making reasonable enquiries and which it is reasonable for the
planning supervisor to expect the principal contractor to need in order
for him to comply with the requirement imposed on him by paragraph
(4); and

(f) such information as the planning supervisor knows or could ascertain
by making reasonable enquiries and which it would be reasonable for
any contractor to know in order to understand how he can comply
with any requirements placed upon him in respect of welfare by or
under the relevant statutory provisions.

(4) The principal contractor shall take such measures as it is reasonable
for a person in his position to take to ensure that the health and safety
plan contains until the end of the construction phase the following
features:

(a) arrangements for the project (including, where necessary, for management of construction work and monitoring of compliance with the
relevant statutory provisions) which will ensure, so far as is reasonably
practicable, the health and safety of all persons at work carrying out
the construction work and all persons who may be affected by the
work of such persons at work, taking account of –

(i) risks involved in the construction work,

(ii) any activity specified in paragraph (5); and

(b) sufficient information about arrangements for the welfare of persons
at work by virtue of the project to enable any contractor to understand
how he can comply with any requirements placed upon him in respect
of welfare by or under the relevant statutory provisions.

(5) An activity is an activity referred to in paragraph (4)(a)(ii) if –

(a) it is an activity of persons at work; and

(b) it is carried out in or on the premises where construction work is or
will be carried out; and

(c) either –

(i) the activity may affect the health or safety of persons at work
carrying out the construction work or persons who may be
affected by the work of such persons at work, or

(ii) the health or safety of the persons at work carrying out the activity may be affected by the work of persons at work carrying out the construction work.

Requirements on and powers of principal contractor

16.—(1) The principal contractor appointed for any project shall –

(a) take reasonable steps to ensure co-operation between all contractors (whether they are sharing the construction site for the purposes of regulation 9 of the Management of Health and Safety at Work Regulations 1992 or otherwise) so far as is necessary to enable each of those contractors to comply with the requirements and prohibitions imposed on him by or under the relevant statutory provisions relating to the construction work;

(b) ensure, so far as is reasonably practicable, that every contractor, and every employee at work in connection with the project complies with any rules contained in the health and safety plan;

(c) take reasonable steps to ensure that only authorised persons are allowed into any premises or part of premises where construction work is being carried out;

(d) ensure that the particulars required to be in any notice given under regulation 7 are displayed in a readable condition in a position where they can be read by any person at work on construction work in connection with the project; and

(e) promptly provide the planning supervisor with any information which –

(i) is in the possession of the principal contractor or which he could ascertain by making reasonable enquiries of a contractor, and

(ii) it is reasonable to believe the planning supervisor would include in the health and safety file in order to comply with the requirements imposed on him in respect thereof in regulation 14, and

(iii) is not in the possession of the planning supervisor.

(2) The principal contractor may –

(a) give reasonable directions to any contractor so far as is necessary to enable the principal contractor to comply with his duties under these Regulations;

(b) include in the health and safety plan rules for the management of the construction work which are reasonably required for the purposes of health and safety.

(3) Any rules contained in the health and safety plan shall be in writing and shall be brought to the attention of persons who may be affected by them.

Information and training

17.—(1) The principal contractor appointed for any project shall ensure, so far as is reasonably practicable, that every contractor is provided with comprehensible information on the risks to the health or safety of that contractor or of any employees or other persons under the control of that contractor arising out of or in connection with the construction work.

(2) The principal contractor shall ensure, so far as is reasonably practicable, that every contractor who is an employer provides any of his employees at work carrying out the construction work with –

(a) any information which the employer is required to provide to those employees in respect of that work by virtue of regulation 8 of the Management of Health and Safety at Work Regulations 1992; and

(b) any health and safety training which the employer is required to provide to those employees in respect of that work by virtue of regulation 11(2)(b) of the Management of Health and Safety at Work Regulations 1992.

Advice from, and views of, persons at work

18. The principal contractor shall –

(a) ensure that employees and self-employed persons at work on the construction work are able to discuss, and offer advice to him on, matters connected with the project which it can reasonably be foreseen will affect their health or safety; and

(b) ensure that there are arrangements for the co-ordination of the views of employees at work on construction work, or of their representatives, where necessary for reasons of health and safety having regard to the nature of the construction work and the size of the premises where the construction work is carried out.

Requirements and prohibitions on contractors

19.—(1) Every contractor shall, in relation to the project –

(a) co-operate with the principal contractor so far as is necessary to enable each of them to comply with his duties under the relevant statutory provisions;

(b) so far as is reasonably practicable, promptly provide the principal contractor with any information (including any relevant part of any risk assessment in his possession or control made by virtue of the Management of Health and Safety at Work Regulations 1992) which might affect the health or safety of any person at work carrying out the construction work or of any person who may be affected by the work of such a person at work or which might justify a review of the health and safety plan;

(c) comply with any directions of the principal contractor given to him under regulation 16(2)(a);

(d) comply with any rules applicable to him in the health and safety plan;

(e) promptly provide the principal contractor with the information in relation to any death, injury, condition or dangerous occurrence which the contractor is required to notify or report by virtue of the Reporting of Injuries, Diseases and Dangerous Occurrences Regulations 1985; and

(f) promptly provide the principal contractor with any information which –

    (i) is in the possession of the contractor or which he could ascertain by making reasonable enquiries of persons under his control, and

    (ii) it is reasonable to believe the principal contractor would provide to the planning supervisor in order to comply with the requirements imposed on the principal contractor in respect thereof by regulation 16(1)(e), and

    (iii) which is not in the possession of the principal contractor.

(2) No employer shall cause or permit any employee of his to work on construction work unless the employer has been provided with the information mentioned in paragraph (4).

(3) No self-employed person shall work on construction work unless he has been provided with the information mentioned in paragraph (4).

(4) The information referred to in paragraphs (2) and (3) is –

(a) the name of the planning supervisor for the project;

(b) the name of the principal contractor for the project; and

(c) the contents of the health and safety plan or such part of it as is relevant to the construction work which any such employee or, as the case may be, which the self-employed person, is to carry out.

(5) It shall be a defence in any proceedings for contravention of paragraph (2) or (3) for the employer or self-employed person to show that he made all reasonable enquiries and reasonably believed –

(a) that he had been provided with the information mentioned in paragraph (4); or

(b) that, by virtue of any provision in regulation 3, this regulation did not apply to the construction work.

Extension outside Great Britain

20. These Regulations shall apply to any activity to which sections 1 to 59 and 80 to 82 of the Health and Safety at Work etc. Act 1974 apply by virtue of article 7 of the Health and Safety at Work etc. Act 1974 (Application outside Great Britain) Order 1989 other than the activities specified in sub-paragraphs (b), (c) and (d) of that article as they apply to any such activity in Great Britain.

Exclusion of civil liability

21. Breach of a duty imposed by these Regulations, other than those imposed by regulation 10 and regulation 16(1)(c), shall not confer a right of action in any civil proceedings.

Enforcement

22. Notwithstanding regulation 3 of the Health and Safety (Enforcing Authority) Regulations 1989, the enforcing authority for these Regulations shall be the Executive.

Transitional provisions

23. Schedule 2 shall have effect with respect to projects which have started, but the construction phase of which has not ended, when these Regulations come into force.

Repeals, revocations and modifications

24.—(1) Subsections (6) and (7) of section 127 of the Factories Act 1961 are repealed.

(2) Regulations 5 and 6 of the Construction (General Provisions) Regulations 1961 are revoked.
(3) The Construction (Notice of Operations and Works) Order 1965 is revoked.
(4) For item (i) of paragraph 4(a) of Schedule 2 to the Health and Safety (Enforcing Authority) Regulations 1989, the following item shall be substituted –

'(i) regulation 7(1) of the Construction (Design and Management) Regulations 1994 (S.I.1994/3140) (which requires projects which include or are intended to include construction work to be notified to the Executive) applies to the project which includes the work; or'.

Signed by order of the Secretary of State.

Phillip Oppenheim

Parliamentary Under Secretary of State, Department of Employment.

19th December 1994

Construction (Health,
Safety and Welfare)
Regulations 1996

Citation and commencement

1. These Regulations may be cited as the Construction (Health, Safety and Welfare) Regulations 1996 and shall come into force on 2nd September 1996.

Interpretation

2.—(1) In these Regulations, unless the context otherwise requires—

'construction site' means any place where the principal work activity being carried out is construction work;

'construction work' means the carrying out of any building, civil engineering or engineering construction work and includes any of the following—

(a) the construction, alteration, conversion, fitting out, commissioning, renovation, repair, upkeep, redecoration or other maintenance (including cleaning which involves the use of water or an abrasive at high pressure or the use of substances classified as corrosive or toxic for the purposes of regulation 5 of the Carriage of Dangerous Goods by Road and Rail (Classification, Packaging and Labelling) Regulations 1994), de-commissioning, demolition or dismantling of a structure,

(b) the preparation for an intended structure, including site clearance, exploration, investigation (but not site survey) and excavation, and laying or installing the foundations of the structure,

(c) the assembly of prefabricated elements to form a structure or the disassembly of prefabricated elements which, immediately before such disassembly, formed a structure,

(d) the removal of a structure or part of a structure or of any product or waste resulting from demolition or dismantling of a structure or from disassembly of prefabricated elements which, immediately before such disassembly, formed a structure, and

(e) the installation, commissioning, maintenance, repair or removal of mechanical, electrical, gas, compressed air, hydraulic, tele-communications, computer or similar services which are normally fixed within or to a structure,

but does not include the exploration for or extraction of mineral resources or activities preparatory thereto carried out at a place where such exploration or extraction is carried out;

'excavation' includes any earthwork, trench, well, shaft, tunnel or underground working;

'fragile material' means any material which would be liable to fail if the weight of any person likely to pass across or work on that material (including the weight of anything for the time being supported or carried by that person) were to be applied to it;

'loading bay' means any facility for loading or unloading equipment or materials for use in construction work;

'personal suspension equipment' means suspended access equipment (other than a working platform) for use by an individual and includes a boatswain's chair and abseiling equipment but it does not include a suspended scaffold or cradle;

'place of work' means any place which is used by any person at work for the purposes of construction work or for the purposes of any activity arising out of or in connection with construction work;

'plant and equipment' includes any machinery, apparatus, appliance or other similar device, or any part thereof, used for the purposes of construction work and any vehicle being used for such purpose;

'structure' means—

(a) any building, steel or reinforced concrete structure (not being a building), railway line or siding, tramway line, dock, harbour, inland navigation, tunnel, shaft, bridge, viaduct, waterworks, reservoir, pipe or pipe-line (whatever, in either case, it contains or is intended to contain), cable, aqueduct, sewer, sewage works, gasholder, road, airfield, sea defence works, river works, drainage works, earthworks, lagoon, dam, wall, caisson, mast, tower, pylon, underground tank, earth retaining structure, or structure designed to preserve or alter any natural feature, and any other structure similar to the foregoing, or

(b) any formwork, falsework, scaffold or other structure designed or used to provide support or means of access during construction work, or

(c) any fixed plant in respect of work which is installation, commissioning, de-commissioning or dismantling and where any such work involves a risk of a person falling more than 2 metres.

'traffic route' means any route the purpose of which is to permit the access to or egress from any part of a construction site for any pedestrians or vehicles, or both, and includes any doorway, gateway, loading bay or ramp;

'vehicle' includes any mobile plant and locomotive and any vehicle towed by another vehicle;

'working platform' means any platform used as a place of work or as a means of access to or egress from that place and includes any scaffold, suspended scaffold, cradle, mobile platform, trestle, gangway, run, gantry, stairway and crawling ladder.

(2) Unless the context otherwise requires, any reference in these Regulations to—

(a) a numbered regulation or Schedule is a reference to the regulation or Schedule in these Regulations so numbered; and
(b) a numbered paragraph is a reference to the paragraph so numbered in the regulation or Schedule in which the reference appears.

Application

3.—(1) Subject to the following paragraphs of this regulation, these Regulations apply to and in relation to construction work carried out by a person at work.

(2) These Regulations shall not apply to any workplace on a construction site which is set aside for purposes other than construction work.
(3) Regulations 15, 19, 20, 21, 22 and 26(1) and (2) apply only to and in relation to construction work carried out by a person at work at a construction site.

Persons upon whom duties are imposed by these Regulations

4.—(1) Subject to paragraph (5), it shall be the duty of every employer whose employees are carrying out construction work and every self-employed person carrying out construction work to comply with the provisions of these Regulations insofar as they affect him or any person at work under his control or relate to matters which are within his control.

(2) It shall be the duty of every person (other than a person having a duty under paragraph (1) or (3)) who controls the way in which any construction work is carried out by a person at work to comply with the provisions of these Regulations insofar as they relate to matters which are within his control.
(3) Subject to paragraph (5), it shall be the duty of every employee carrying out construction work to comply with the requirements of these Regulations insofar as they relate to the performance of or the refraining from an act by him.

(4)   It shall be the duty of every person at work—

    (a)   as regards any duty or requirement imposed on any other person under these Regulations, to co-operate with that person so far as is necessary to enable that duty or requirement to be performed or complied with; and

    (b)   where working under the control of another person, to report to that person any defect which he is aware may endanger the health or safety of himself or another person.

(5)   This regulation shall not apply to regulations 22 and 29(2), which expressly say on whom the duties are imposed.

Safe places of work

5.—(1)   There shall, so far as is reasonably practicable, be suitable and sufficient safe access to and egress from every place of work and to any other place provided for the use of any person while at work, which access and egress shall be without risks to health and properly maintained.

(2)   Every place of work shall, so far as is reasonably practicable, be made and kept safe for, and without risks to health to, any person at work there.

(3)   Suitable and sufficient steps shall be taken to ensure, so far as is reasonably practicable, that no person gains access to any place which does not comply with the requirements of paragraphs (1) or (2).

(4)   Paragraphs (1) to (3) shall not apply in relation to a person engaged in work for the purpose of making any place safe, provided all practicable steps are taken to ensure the safety of that person whilst engaged in that work.

(5)   Every place of work shall, so far as is reasonably practicable and having regard to the nature of the work being carried out there, have sufficient working space and be so arranged that it is suitable for any person who is working or who is likely to work there.

Falls

6.—(1)   Suitable and sufficient steps shall be taken to prevent, so far as is reasonably practicable, any person falling.

(2)   In any case where the steps referred to in paragraph (1) include the provision of—

    (a)   any guard-rail, toe-board, barrier or other similar means of protection; or

    (b)   any working platform,

it shall comply with the provisions of Schedule 1 and Schedule 2 respectively.

(3) Without prejudice to the generality of paragraph (1) and subject to paragraph (6), where any person is to carry out work at a place from which he is liable to fall a distance of 2 metres or more or where any person is to use a means of access to or egress from a place of work from which access or egress he is liable to fall a distance of 2 metres or more—

(a) there shall, subject to sub-paragraphs (c) and (d) below and paragraph (9), be provided and used suitable and sufficient guard-rails and toe-boards, barriers or other similar means of protection to prevent, so far as is reasonably practicable, the fall of any person from that place, which guard-rails, toe-boards, barriers and other similar means of protection shall comply with the provisions of Schedule 1; and

(b) where it is necessary in the interest of the safety of any person that a working platform be provided, there shall, subject to sub-paragraphs (c) and (d) below, be so provided and used a sufficient number of working platforms which shall comply with the provisions of Schedule 2; and

(c) where it is not practicable to comply with all or any of the requirements of sub-paragraphs (a) or (b) above or where due to the nature or the short duration of the work compliance with such requirements is not reasonably practicable, there shall, subject to sub-paragraph (d) below, be provided and used suitable personal suspension equipment which shall comply with the provisions of Schedule 3; and

(d) where it is not practicable to comply with all or any of the requirements of sub-paragraphs (a), (b) or (c) above or where due to the nature or the short duration of the work compliance with such requirements is not reasonably practicable, such requirements of those sub-paragraphs as can be complied with shall be complied with and, in addition, there shall be provided and used suitable and sufficient means for arresting the fall of any person which shall comply with the provisions of Schedule 4.

(4) Means for the prevention of, or for protection from, falls provided pursuant to sub-paragraph (a) and (d) of paragraph (3) may be removed for the time and to the extent necessary for the movement of materials, but shall be replaced as soon as practicable.

(5) A ladder shall not be used as, or as a means of access to or egress from, a place of work unless it it reasonable to do so having regard to—

(a) the nature of the work being carried out and its duration; and
(b) the risks to the safety to any person arising from the use of the ladder.

(6) Where a ladder is used pursuant to paragraph (5)—

(a) it shall comply with the provisions of Schedule 5; and
(b) the provisions of paragraph (3) shall not apply.

(7) Any equipment provided pursuant to this regulation shall be properly maintained.

(8)

    (a) The installation or erection of any scaffold provided pursuant to paragraph (1) or sub-paragraph (b) of paragraph (3) and any substantial addition or alteration to such scaffold shall be carried out only under the supervision of a competent person.

    (b) The installation or erection of any personal suspension equipment or any means of arresting falls provided pursuant to sub-paragraphs (c) or (d) of paragraph (3) shall be carried out only under the supervision of a competent person, and for the purposes of this paragraph installation shall not include the personal attachment of any equipment or means of preventing falls to the person for whose safety such equipment or means is provided.

(9) No toe-boards shall be required in respect of any stairway, or any rest platform forming part of a scaffold, where such stairway or platform is used solely as a means of access to or egress from any place of work, provided that the stairway or platform is not being used for the keeping or storing of any material or substance.

Fragile material

7.—(1) Suitable and sufficient steps shall be taken to prevent any person from falling through any fragile material.

(2) Without prejudice to the generality of paragraph (1),

    (a) no person shall pass across, or work on or from, fragile material through which he would be liable to fall 2 metres or more unless suitable and sufficient platforms, coverings or other similar means of support are provided and used so that the weight of any person so passing or working is supported by such supports; and

    (b) no person shall pass or work near fragile material through which he would be liable to fall 2 metres or more unless there are provided suitable and sufficient guard-rails, coverings or other similar means for preventing, so far as is reasonably practicable, any person so passing or working from falling through that material; and

    (c) where any person may pass across or near or work on or near fragile material through which, were it not for the provisions of sub-paragraphs (a) and (b) above, he would be liable to fall 2 metres or more, prominent warning notices shall be affixed at the approach to the place where the material is situated.

Falling objects

8.—(1) Where necessary to prevent danger to any person, suitable and sufficient steps shall be taken to prevent, so far as is reasonably practicable, the fall of any material or object.

(2) In any case where the steps referred to in paragraph (1) include the provision of—

    (a) any guard-rail, toe-board, barrier or other similar means of protection; or

    (b) any working platform,

it shall comply with the provisions of Schedule 1 and Schedule 2 respectively.

(3) Where it is not reasonably practicable to comply with the requirements of paragraph (1) or where it is otherwise necessary in the interests of safety, suitable and sufficient steps shall be taken to prevent any person from being struck by any falling material or object which is liable to cause injury.

(4) No material or object shall be thrown or tipped from a height in circumstances where it is liable to cause injury to any person.

(5) Materials and equipment shall be stored in such a way as to prevent danger to any person arising from the collapse, overturning or unintentional movement of such materials or equipment.

Stability of structures

9.—(1) All practicable steps shall be taken, where necessary to prevent danger to any person, to ensure that any new or existing structure or any part of such structure which may become unstable or in a temporary state of weakness or instability due to the carrying out of construction work (including any excavation work) does not collapse accidentally.

(2) No part of a structure shall be so loaded as to render it unsafe to any person.

(3) Any buttress, temporary support or temporary structure used to support a permanent structure pursuant to paragraph (1) shall be erected or dismantled only under the supervision of a competent person.

Demolition or dismantling

10.—(1) Suitable and sufficient steps shall be taken to ensure that the demolition or dismantling of any structure, or any part of any structure, being demolition or dismantling which gives rise to a risk of danger to any person, is planned and carried out in such a manner as to prevent, so far as is practicable, such danger.

(2) Demolition or dismantling to which paragraph (1) applies shall be planned and carried out only under the supervision of a competent person.

Explosives

11. An explosive charge shall be used or fired only if suitable and sufficient steps have been taken to ensure that no person is exposed to risk of injury from the explosion or from projected or flying material caused thereby.

Excavations

12.—(1) All practicable steps shall be taken, where necessary to prevent danger to any person, to ensure that any new or existing excavation or any part of such excavation which may be in a temporary state of weakness or instability due to the carrying out of construction work (including other excavation work) does not collapse accidentally.

(2) Suitable and sufficient steps shall be taken to prevent, so far as is reasonably practicable, any person from being buried or trapped by a fall or dislodgement of any material.

(3) Without prejudice to the generality of paragraph (2), where it is necessary for the purpose of preventing any danger to any person from a fall or dislodgement of any material from a side or the roof of or adjacent to any excavation, that excavation shall as early as practicable in the course of the work be sufficiently supported so as to prevent, so far as is reasonably practicable, the fall or dislodgement of such material.

(4) Suitable and sufficient equipment for supporting an excavation shall be provided to ensure that the requirements of paragraphs (1) to (3) may be complied with.

(5) The installation, alteration or dismantling of any support for an excavation pursuant to paragraphs (1), (2) or (3) shall be carried out only under the supervision of a competent person.

(6) Where necessary to prevent danger to any person, suitable and sufficient steps shall be taken to prevent any person, vehicle or plant and equipment, or any accumulation of earth or other material, from falling into any excavation.

(7) Where a collapse of an excavation would endanger any person, no material, vehicle or plant and equipment shall be placed or moved near any excavation where it is likely to cause such collapse.

(8) No excavation work shall be carried out unless suitable and sufficient steps have been taken to identify and, so far as is reasonably practicable, prevent any risk of injury arising from any underground cable or other underground service.

Cofferdams and caissons

13.—(1) Every cofferdam or caisson and every part thereof shall be of suitable design and construction, of suitable and sound material and of sufficient strength and capacity for the purpose for which it is used, and shall be properly maintained.

(2) The construction, installation, alteration or dismantling of a cofferdam or caisson shall take place only under the supervision of a competent person.

Prevention of drowning

14.—(1) Where during the course of construction work any person is liable to fall into water or other liquid with a risk of drowning, suitable and sufficient steps shall be taken—

(a) to prevent, so far as is reasonably practicable, such person from so falling; and

(b) to minimise the risk of drowning in the event of such a fall; and

(c) to ensure that suitable rescue equipment is provided, maintained and, when necessary, used so that such person may be promptly rescued in the event of such a fall.

(2) Suitable and sufficient steps shall be taken to ensure the safe transport of any person conveyed by water to or from any place of work.

(3) Any vessel used to convey any person by water to or from a place of work—

(a) shall be of suitable construction; and

(b) shall be properly maintained; and

(c) shall be under the control of a competent person; and

(d) shall not be overcrowded or overloaded.

Traffic routes

15.—(1) Every construction site shall be organised in such a way that, so far as is reasonably practicable, pedestrians and vehicles can move safely and without risks to health.

(2) Traffic routes shall be suitable for the persons or vehicles using them, sufficient in number, in suitable positions and of sufficient size.

(3) Without prejudice to the generality of paragraph (2), traffic routes shall not satisfy the requirements of that paragraph unless suitable and sufficient steps are taken to ensure that—

(a) pedestrians or, as the case may be, vehicles may use a traffic route without causing danger to the health or safety of persons near it;

(b) any door or gate used or intended to be used by pedestrians and which leads onto a traffic route for vehicles is sufficiently separated from that traffic route to enable pedestrians from a place of safety to see any approaching vehicle or plant;

(c) there is sufficient separation between vehicles and pedestrians to ensure safety or, where this is not reasonably practicable—

    (i) there are provided other means for the protection of pedestrians; and

    (ii) there are effective arrangements for warning any person liable to be crushed or trapped by any vehicle of the approach of that vehicle;

(d) any loading bay has at least one exit point for the exclusive use of pedestrians; and

(e) where it is unsafe for pedestrians to use any gate intended primarily for vehicles, one or more doors for pedestrians is provided in the immediate vicinity of any such gate, which door shall be clearly marked and kept free from obstruction.

(4) No vehicle shall be driven on a traffic route unless, so far as is reasonably practicable, that traffic route is free from obstruction and permits sufficient clearance.

(5) Where it is not reasonably practicable to comply with all or any of the requirements of paragraph (4), suitable and sufficient steps shall be taken to warn the driver of the vehicle and any other person riding thereon of any approaching obstruction or lack of clearance.

(6) Every traffic route shall be indicated by suitable signs where necessary for reasons of health or safety.

Doors and gates

16.—(1) Where necessary to prevent the risk of injury to any person, any door, gate or hatch (including a temporary door, gate or hatch) shall incorporate or be fitted with suitable safety devices.

(2) Without prejudice to the generality of paragraph (1), a door, gate or hatch shall not comply with that paragraph unless—

(a) any sliding door, gate or hatch has a device to prevent it coming off its track during use;

(b) any upward opening door, gate or hatch has a device to prevent it falling back;

(c) any powered door, gate or hatch has suitable and effective features to prevent it causing injury by trapping any person;

(d) where necessary for reasons of health or safety, any powered door, gate or hatch can be operated manually unless it opens automatically if the power fails.

(3) This regulation shall not apply to any door, gate or hatch forming part of any mobile plant and equipment.

Vehicles

17.—(1) Suitable and sufficient steps shall be taken to prevent or control the unintended movement of any vehicle.

(2) Suitable and sufficient steps shall be taken to ensure that, where any person may be endangered by the movement of any vehicle, the person having effective control of the vehicle shall give warning to any person who is liable to be at risk from the movement of the vehicle.

(3) Any vehicle being used for the purposes of construction work shall when being driven, operated or towed—

   (a) be driven, operated or towed in such a manner as is safe in the circumstances; and
   (b) be loaded in such a way that it can be driven, operated or towed safely.

(4) No person shall ride or be required or permitted to ride on any vehicle being used for the purposes of construction work otherwise than in a safe place thereon provided for that purpose.

(5) No person shall remain or be required or permitted to remain on any vehicle during the loading or unloading of any loose material unless a safe place of work is provided and maintained for such person.

(6) Where any vehicle is used for excavating or handling (including tipping) materials, suitable and sufficient measures shall be taken so as to prevent such vehicle from falling into any excavation or pit, or into water, or overrunning the edge of any embankment or earthwork.

(7) Suitable plant and equipment shall be provided and used for replacing on its track or otherwise safely moving any rail vehicle which may become derailed.

Prevention of risk from fire etc.

18.—

Suitable and sufficient steps shall be taken to prevent, so far as is reasonably practicable, the risk of injury to any person during the carrying out of construction work arising from—

   (a) fire or explosion;
   (b) flooding; or
   (c) any substance liable to cause asphyxiation.

Emergency routes and exits

19.—(1) Where necessary in the interests of the health and safety of any person on a construction site, a sufficient number of suitable emergency routes and exits shall be provided to enable any person to reach a place of safety quickly in the event of danger.

(2) An emergency route or exit provided pursuant to paragraph (1) shall lead as directly as possible to an identified safe area.

(3) Any emergency route and exit provided in accordance with paragraph (1), and any traffic route or door giving access thereto, shall be kept clear and free from obstruction, and, where necessary, provided with emergency lighting so that such emergency route or exit may be used at any time.

(4) Any provision for emergency routes and exits made under paragraph (1) shall have regard to—

(a) the type of work for which the construction site is being used;

(b) the characteristics and size of the construction site and the number and location of places of work on that site;

(c) the plant and equipment being used;

(d) the number of persons likely to be present on the site at any one time; and

(e) the physical and chemical properties of any substances or materials on or likely to be on the site.

(5) All emergency routes or exits shall be indicated by suitable signs.

Emergency procedures

20.—(1) Where necessary in the interests of the health and safety of any person on a construction site, there shall be prepared and, when necessary, implemented suitable and sufficient arrangements for dealing with any forseeable emergency, which arrangements shall include procedures for any necessary evacuation of the site or any part thereof.

(2) Without prejudice to the generality of paragraph (1), arrangements prepared pursuant to that paragraph shall have regard to those matters set out in paragraph (4) of regulation 19.

(3) Where arrangements are prepared pursuant to paragraph (1), suitable and sufficient steps shall be taken to ensure that—

(a) every person to whom the arrangements extend is familiar with those arrangements; and

(b) the arrangements are tested by being put into effect at suitable intervals.

Fire detection and fire-fighting

21.—(1) Without prejudice to the provisions of any other enactment, there shall be provided on a construction site where necessary in the interests of the health and safety of any person at work on that site—

    (a) suitable and sufficient fire-fighting equipment; and

    (b) suitable and sufficient fire detectors and alarm systems,

which shall be suitably located.

(2) Any provision for fire-fighting equipment, fire detectors and alarm systems made under paragraph (1) shall have regard to those matters set out in paragraph (4) of regulation 19.

(3) Any fire-fighting equipment, fire detector or alarm system provided under paragraph (1) shall be properly maintained and subject to examination and testing at such intervals as to ensure that such equipment, detector or system remains effective.

(4) Any fire-fighting equipment which is not designed to come into use automatically shall be easily accessible.

(5) Every person at work on a construction site shall, so far as is reasonably practicable, be instructed in the correct use of any fire-fighting equipment which it may be necessary for him to use.

(6) Where a work activity may give rise to a particular risk of fire, a person shall not carry out such work unless he is suitably instructed so as to prevent, so far as is reasonably practicable, that risk.

(7) Fire-fighting equipment shall be indicated by suitable signs.

Welfare facilities

22.—(1) It shall be the duty of any person in control of a construction site to ensure, so far as is reasonably practicable, that the requirements of this regulation are complied with in relation to that site.

(2) It shall be the duty of every employer and every self-employed person to ensure that the provisions of paragraphs (3) to (8) are complied with in respect of any person at work on a construction site who is under his control.

(3) Suitable and sufficient sanitary conveniences shall be provided or made available at readily accessible places, which conveniences shall, so far as is reasonably practicable, comply with the provisions of paragraphs 1 to 3 of Schedule 6.

(4) Suitable and sufficient washing facilities, including showers if required by the nature of the work or for health reasons, shall be provided or made available at readily accessible places, which facilities shall, so far as is reasonably practicable, comply with the provisions of paragraphs 4 to 9 of

Schedule 6, save that in respect of the provision of showers, paragraph 4(a) of that Schedule shall not apply.

(5) An adequate supply of wholesome drinking water shall be provided or made available at readily accessible and suitable places, which supply shall, so far as is reasonably practicable, comply with the provisions of paragraphs 10 and 11 of Schedule 6.

(6) Suitable and sufficient accommodation shall be provided or made available—

(a) for the clothing of any person at work on a construction site and which is not worn during working hours; and

(b) for special clothing which is worn by any person at work on a construction site but which is not taken home,

which accommodation shall, so far as is reasonably practicable, comply with the provisions of paragraph 12 of Schedule 6.

(7) Suitable and sufficient facilities shall be provided or made available to change clothing in all cases where—

(a) a person has to wear special clothing for the purpose of his work; and

(b) that person cannot, for reasons of health or propriety, be expected to change elsewhere,

which facilities shall, so far as is reasonably practicable, comply with the provisions set out in paragraph 13 of Schedule 6.

(8) Suitable and sufficient facilities for rest shall be provided or made available at readily accessible places, which facilities shall, so far as is reasonably practicable, comply with the provisions of paragraph 14 of Schedule 6.

Fresh air

23.—(1) Suitable and sufficient steps shall be taken to ensure, so far as is reasonably practicable, that every workplace or approach thereto has sufficient fresh or purified air to ensure that the place or approach is safe and without risks to health.

(2) Any plant used for the purpose of complying with paragraph (1) shall, where necessary for reasons of health or safety, include an effective device to give visible or audible warning of any failure of the plant.

Temperature and weather protection

24.—(1) Suitable and sufficient steps shall be taken to ensure, so far as is reasonably practicable, that during working hours the temperature at any indoor place of work to which these Regulations apply is reasonable having regard to the purpose for which that place is used.

(2) Every place of work outdoors shall, where necessary to ensure the health and safety of persons at work there, be so arranged that, so far as is reasonably practicable and having regard to the purpose for which that place is used and any protective clothing or equipment provided for the use of any person at work there, it provides protection from adverse weather.

Lighting

25.—(1) There shall be suitable and sufficient lighting in respect of every place of work and approach thereto and every traffic route, which lighting shall, so far as is reasonably practicable, be by natural light.

(2) The colour of any artificial lighting provided shall not adversely affect or change the perception of any sign or signal provided for the purposes of health and safety.

(3) Without prejudice to the generality of paragraph (1), suitable and sufficient secondary lighting shall be provided in any place where there would be a risk to the health or safety of any person in the event of failure of primary artificial lighting.

Good order

26.—(1) Every part of a construction site shall, so far as is reasonably practicable, be kept in good order and every part of a construction site which is used as a place of work shall be kept in a reasonable state of cleanliness.

(2) Where necessary in the interests of health and safety, the perimeter of a construction site shall, so far as is reasonably practicable, be identified by suitable signs and the site shall be so arranged that its extent is readily identifiable.

(3) No timber or other material with projecting nails shall –

(a) be used in any work in which the nails may be a source of danger to any person; or

(b) be allowed to remain in any place where the nails may be a source of danger to any person.

Plant and equipment

27.—(1) All plant and equipment used for the purpose of carrying out construction work shall, so far as is reasonably practicable, be safe and without risks to health and shall be of good construction, of suitable and sound materials and of sufficient strength and suitability for the purpose for which it is used or provided.

(2) All plant and equipment used for the purpose of carrying out construction work shall be used in such a manner and be maintained in such condition that, so far as is reasonably practicable, it remains safe and without risks to health at all times when it is being used.

Training

28. Any person who carries out any activity involving construction work where training, technical knowledge or experience is necessary to reduce the risks of injury to any person shall possess such training, knowledge or experience, or be under such degree of supervision by a person having such training, knowledge or experience, as may be appropriate having regard to the nature of the activity.

Inspection

29.—(1) Subject to paragraph (2), a place of work referred to in column 1 of Schedule 7 shall be used to carry out construction work only if that place has been inspected by a competent person at the times set out in the corresponding entry in column 2 of that Schedule and the person who has carried out the inspection is satisfied that the work can be safely carried out at that place.

(2) Without prejudice to paragraph (1), where the place of work is a part of a scaffold, excavation, cofferdam or caisson, any employer or any other person who controls the way in which construction work is carried out by persons using that part shall ensure that the scaffold, excavation, cofferdam or caisson is stable and of sound construction and that the safeguards required by these Regulations are in place before his employees or persons under his control first use that place of work.

(3) Where the person who has carried out an inspection pursuant to paragraph (1) is not satisfied that construction work can safely be carried out at that place –

(a) where the inspection was carried out on behalf of another person, he shall inform that person of any matters about which he is not satisfied; and

(b) the place of work shall not be used until the matters identified have been satisfactorily remedied.

(4) An inspection of a place of work carried out pursuant to paragraph (1) shall include an inspection of any plant and equipment and any materials which affect the safety of that place of work.

Reports

30.—(1) Subject to paragraphs (5) and (6), where an inspection is required under regulation 29(1), the person who carries out such inspection shall, before the end of the working period within which the inspection is completed, prepare a report which shall include the particulars set out in Schedule 8.

(2) A person who prepares a report under paragraph (1) shall, within 24 hours of completing the inspection to which the report relates, provide the report or a copy thereof to the person on whose behalf the inspection was carried out.

(3) The report or a copy thereof prepared for the purposes of paragraph (1) shall be kept at the site of the place of work in respect of which the inspection was carried out and, after that work is completed, shall be retained at an office of the person on whose behalf the inspection was carried out for a period of 3 months from the date of such completion.

(4) A report prepared for the purposes of paragraph (1) shall at all reasonable times be open to inspection by any inspector, and the person keeping such report shall send to any such inspector such extracts therefrom or copies thereof as the inspector may from time to time require for the purpose of the execution of his duties.

(5) No report is required to be prepared under paragraph (1) in respect of any working platform or alternative means of support from no part of which a person is liable to fall more than 2 metres.

(6) Nothing in this regulation shall require –

(a) a report to be prepared in respect of any mobile tower scaffold unless it remains erected in the same place for a period of 7 days or more;

(b) as regards an inspection carried out on a place of work for the purposes of paragraph 1(ii) of column 2 of Schedule 7, the preparation of more than one report on that place within any period of 24 hours; or

(c) as regards an inspection carried out on a place of work for the purposes of paragraph 2(i) or 3(i) of column 2 of Schedule 7, the preparation of more than one report on that place within any period of 7 days.

Exemption certificates

31.—(1) Subject to paragraph (2), the Executive may, by a certificate in writing, exempt –

(a) any person or class of person;

(b) any premises or class of premises; or

(c) any plant and equipment,

from any requirement or prohibition imposed by these Regulations and any such exemption may be granted subject to conditions and to a limit of time and may be revoked at any time by a certificate in writing.

(2) The Executive shall not grant any such exemption unless, having regard to the circumstances of the case and in particular to –

(a) the conditions, if any, which it proposes to attach to the exemption; and

(b) any other requirements imposed by or under any enactments which apply to the case,

it is satisfied that the health and safety of persons who are likely to be affected by the exemption will not be prejudiced in consequence of it.

Extension outside Great Britain

32. These Regulations shall apply to any activity to which sections 1 to 59 and 80 to 82 of the Health and Safety at Work etc. Act 1974 apply by virtue of article 8 of the Health and Safety at Work etc. Act (Application outside Great Britain) Order 1995 other than the activities specified in sub-paragraphs (b), (c), (d) and (e) of that article as they apply to any such activity in Great Britain.

Enforcement in respect of fire

33.—(1) Subject to paragraph (2), the fire authority within the meaning of section 43(1) of the Fire Precautions Act 1971 shall be the enforcing authority as regards –

(a) regulations 19 and 20 insofar as those regulations relate to fire; and

(b) regulation 21,

in respect of a construction site which is contained within, or forms part of, premises which are occupied by persons other than those carrying out the construction work or any activity arising from such work.

(2) Paragraph (1) shall not apply in respect of any premises of a description specified in Part I of Schedule 1 to the Fire Certificates (Special Premises) Regulations 1976.

Modifications

34. The Act and instruments mentioned in Schedule 9 shall be modified to the extent specified in that Schedule.

Revocations

35. The instruments mentioned in column 1 of Schedule 10 are revoked to the extent specified in column 3 of that Schedule.

John Gummer

Secretary of State for the Environment

14th June 1996

Schedule 1, Regulations 6(2), 6(3)(a) and 8(2), Requirements for guard-rails etc.

1. A guard-rail, toe-board, barrier or other similar means of protection shall –

(a) be suitable and of sufficient strength and rigidity for the purpose or purposes for which it is being used; and
(b) be so placed, secured and used as to ensure, so far as is reasonably practicable, that it does not become accidentally displaced.

2. Any structure or any part of a structure which supports a guard-rail, toe-board, barrier or other similar means of protection or to which a guard-rail, toe-board, barrier or other similar means of protection is attached shall be of sufficient strength and suitable for the purpose of such support or attachment.

3. The main guard-rail or other similar means of protection shall be at least 910 millimetres above the edge from which any person is liable to fall.

4. There shall not be an unprotected gap exceeding 470 millimetres between any guard-rail, toe-board, barrier or other similar means of protection.

5. Toe-boards or other similar means of protection shall not be less than 150 millimetres high.

6. Guard-rails, toe-boards, barriers and other similar means of protection shall be so placed as to prevent, so far as is practicable, the fall of any person, or any material or object, from any place of work.

Schedule 2, Regulations 6(2), 6(3)(b) and 8(2), Requirements for working platforms

Interpretation

1. In this Schedule, 'supporting structure' means any structure used for the purpose of supporting a working platform and includes any plant and equipment used for that purpose.

Condition of surfaces

2. Any surface upon which any supporting structure rests shall be stable, of sufficient strength and of suitable composition safely to support the supporting

structure, the working platform and any load intended to be placed on the working platform.

Stability of supporting structure

3. Any supporting structure shall –

(a) be suitable and of sufficient strength and rigidity for the purpose or purposes for which it is being used; and
(b) be so erected and, where necessary, securely attached to another structure as to ensure that it is stable; and
(c) when altered or modified, be so altered or modified as to ensure that it remains stable.

Stability of working platform

4. A working platform shall –

(a) be suitable and of sufficient strength and rigidity for the purpose or purposes for which it is intended to be used or is being used; and
(b) be so erected and used as to ensure, so far as is reasonably practicable, that it does not become accidentally displaced so as to endanger any person; and
(c) when altered or modified, be so altered or modified as to ensure that it remains stable; and
(d) be dismantled in such a way as to prevent accidental displacement.

Safety on working platforms

5. A working platform shall –

(a) be of sufficient dimensions to permit the free passage of persons and the safe use of any equipment or materials required to be used and to provide, so far as is reasonably practicable, a safe working area having regard to the work there being carried out; and
(b) without prejudice to paragraph (a), be not less than 600 millimetres wide; and
(c) be so constructed that the surface of the working platform has no gap giving rise to the risk of injury to any person or, where there is a risk of any person below the platform being struck, through which any material or object could fall; and
(d) be so erected and used, and maintained in such condition, as to prevent, so far as is reasonably practicable –

(i)   the risk of slipping or tripping; or

(ii) any person being caught between the working platform and any adjacent structure; and

(e) be provided with such handholds and footholds as are necessary to prevent, so far as is reasonably practicable, any person slipping from or falling from the working platform.

Loading

6. A working platform and any supporting structure shall not be loaded so as to give rise to a danger of collapse or to any deformation which could affect its safe use.

Schedule 3, Regulation 6(3)(c), Requirements for personal suspension equipment

1. Personal suspension equipment shall be suitable and of sufficient strength for the purpose or purposes for which it is being used having regard to the work being carried out and the load, including any person, it is intended to bear.
2. Personal suspension equipment shall be securely attached to a structure or to plant and the structure or plant and the means of attachment thereto shall be suitable and of sufficient strength and stability for the purpose of supporting that equipment and the load, including any person, it is intended to bear.
3. Suitable and sufficient steps shall be taken to prevent any person falling or slipping from personal suspension equipment.
4. Personal suspension equipment shall be installed or attached in such a way as to prevent uncontrolled movement of that equipment.

Schedule 4, Regulation 6(3)(d), Requirements for means of arresting falls

1. In this Schedule, 'equipment' means any equipment provided for the purpose of arresting the fall of any person at work and includes any net or harness provided for that purpose.
2. The equipment shall be suitable and of sufficient strength to safely arrest the fall of any person who is liable to fall.
3. The equipment shall be securely attached to a structure or to plant and the structure or plant and the means of attachment thereto shall be suitable and of sufficient strength and stability for the purpose of safely supporting the equipment and any person who is liable to fall.
4. Suitable and sufficient steps shall be taken to ensure, so far as practicable, that in the event of a fall by any person the equipment does not itself cause injury to that person.

Schedule 5, Regulation 6(6), Requirements for ladders

1. Any surface upon which a ladder rests shall be stable, level and firm, of sufficient strength and of suitable composition safely to support the ladder and any load intended to be placed on it.

2. A ladder shall –

(a) be suitable and of sufficient strength for the purpose or purposes for which it is being used;

(b) be so erected as to ensure that it does not become displaced; and

(c) where it is of a length when used of 3 metres or more, be secured to the extent that it is practicable to do so and where it is not practicable to secure the ladder a person shall be positioned at the foot of the ladder to prevent it slipping at all times when it is being used.

3. All ladders used as a means of access between places of work shall be sufficiently secured so as to prevent the ladder slipping or falling.

4. The top of any ladder used as a means of access to another level shall, unless a suitable alternative handhold is provided, extend to a sufficient height above the level to which it gives access so as to provide a safe handhold.

5. Where a ladder or run of ladders rises a vertical distance of 9 metres or more above its base, there shall, where practicable, be provided at suitable intervals sufficient safe landing areas or rest platforms.

Schedule 6, Regulation 22, Welfare facilities

Sanitary conveniencies

1. Rooms containing sanitary conveniencies shall be adequately ventilated and lit.

2. Sanitary conveniences and the rooms containing them shall be kept in a clean and orderly condition.

3. Separate rooms containing sanitary conveniences shall be provided for men and women, except where and so far as each convenience is in a separate room the door of which is capable of being secured from the inside.

Washing facilities

4. Washing facilities shall be provided –

(a) in the immediate vicinity of every sanitary convenience, whether or not provided elsewhere; and

(b) in the vicinity of any changing rooms required by paragraph (7) of regulation 22 whether or not provided elsewhere.

5. Washing facilities shall include –

(a) a supply of clean hot and cold, or warm, water (which shall be running water so far as is reasonably practicable); and
(b) soap or other suitable means of cleaning; and
(c) towels or other suitable means of drying.

6. Rooms containing washing facilities shall be sufficiently ventilated and lit.
7. Washing facilities and the rooms containing them shall be kept in a clean and orderly condition.
8. Subject to paragraph 9 below, separate washing facilities shall be provided for men and women, except where and so far as they are provided in a room the door of which is capable of being secured from inside and the facilities in each such room are intended to be used by only one person at a time.
9. Paragraph 8 above shall not apply to facilities which are provided for washing hands, forearms and face only.

Drinking water

10. Every supply of drinking water shall be conspicuously marked by an appropriate sign where necessary for reasons of health and safety.
11. Where a supply of drinking water is provided, there shall also be provided a sufficient number of suitable cups or other drinking vessels unless the supply of drinking water is in a jet from which persons can drink easily.

Accommodation for clothing

12. Accommodation for clothing shall include or allow for facilities for drying clothing.

Facilities for changing clothing

13. The facilities for changing clothing shall be separate facilities for, or separate use of facilities by, men and women where necessary for reasons of propriety.

Facilities for rest

14. Rest facilities shall –

(a) include rest facilities provided in one or more rest rooms or rest areas;

(b) include rest rooms or rest areas with suitable arrangements to protect non-smokers from discomfort caused by tobacco smoke;

(c) where necessary, include suitable facilities for any person at work who is a pregnant woman or nursing mother to rest;

(d) include suitable arrangements to ensure that meals can be prepared and eaten; and

(e) include the means for boiling water.

Schedule 7, Regulation 29(1), Places of work requiring inspection

Column 1	Column 2
Place of Work	*Time of inspection*
1. Any working platform or part thereof or any personal suspension equipment provided pursuant to paragraph (3)(b) or (c) of regulation 6.	1. (i) Before being taken into use for the first time; and (ii) after any substantial addition, dismantling or other alteration; and (iii) after any event likely to have affected its strength or stabililty; and (iv) at regular intervals not exceeding 7 days since the last inspection.
2. Any excavation which is supported pursuant to paragraphs (1), (2) or (3) of regulation 12.	2. (i) Before any person carries out work at the start of every shift; and (ii) after any event likely to have affected the strength or stability of the excavation or any part thereof; and (iii) after any accidental fall of rock or earth or other material.
3. Cofferdams and caissons	3. (i) Before any person carries out work at the start of every shift; and (ii) after any event likely to have affected the strength or stability of the cofferdam or caisson or any part thereof.

Schedule 8, Regulation 30, Particulars to be included in a report of inspection

1. Name and address of the person on whose behalf the inspection was carried out.

2. Location of the place of work inspected.

3. Description of the place of work or part of that place inspected (including any plant and equipment and materials, if any).

4. Date and time of the inspection.

5. Details of any matter identified that could give rise to a risk to the health or safety of any person.

6. Details of any action taken as a result of any matter identified in paragraph 5 above.

7. Details of any further action considered necessary.

8. Name and position of the person making report.

Schedule 9, Regulation 34, Modifications

The Mines and Quarries Act 1954

1. In sub-section (5) of section 184 of the Mines and Quarries Act 1954 –

(a) the words after the semi-colon shall be deleted; and

(b) in place of the semi-colon, there shall be substituted a full stop.

The Factories Act 1961

2. The Factories Act 1961 is modified as follows –

(a) in section 127, sub-section (3) shall cease to have effect;

(b) in section 176, in sub-section (1) –

    (i) the definition of 'building operation' shall be deleted and the following substituted –

' "building operation" and "work of engineering construction" mean "construction work" within the meaning assigned to that phrase by regulation 2(1) of the Construction (Design and Management) Regulations 1994 (S.I. 1994/3140);';

    (ii) the definition of 'work of engineering construction' is deleted.

The Construction (Lifting Operations) Regulations 1961

3. The Construction (Lifting Operations) Regulations 1961 are modified as follows –

(a) in regulation 3 –

    (i) in place of 'regulation 47', substitute 'regulations 47 and 48A';

    (ii) after '48', insert, 48A';

(b) after regulation 48 the following shall be added –

'Suspended scaffolds (not power operated)

48A. –

(1) Without prejudice to any requirement of these Regulations as respects lifting appliances, chains, ropes and lifting gear used in connection therewith, the requirements of this regulation shall be observed as respects –

   (a) every suspended scaffold; and
   (b) plant or equipment which is permanent plant or equipment of a building and which, but for the fact that it is permanently provided, would be a suspended scaffold,

   being in any case a suspended scaffold, plant or equipment which is not raised or lowered by a power-driven lifting appliance or power-driven lifting appliances and no such suspended scaffold, plant or equipment shall be used unless it complies with the requirements of this regulation.

(2) In the application of the succeeding paragraphs of this regulation, references therein to suspended scaffolds shall be construed as references to suspended scaffolds to which this regulation applies and as including references to plant or equipment of the kind referred to in sub-paragraph (b) of the foregoing paragraph of this regulation.

(3) Every suspended scaffold shall be provided with adequate and suitable chains or ropes and winches or other lifting appliances or similar devices and shall be suspended from suitable outriggers, joists, runways, rail tracks or other equally safe anchorage.

(4) The winches or other lifting appliances or similar devices of a suspended scaffold shall be –

   (a) provided with a brake or similar device which comes into operation when the operating handle or lever is released; and
   (b) adequately protected against the effects of weather, dust or material likely to cause damage.

(5) The outriggers for a suspended scaffold shall be of adequate length and strength and properly installed and supported and, subject to paragraph (12) of this regulation, shall be installed horizontally and provided with adequate stops at their outer ends. The outriggers shall be properly spaced having regard to the construction of the scaffold and of the runway, joist or rail track on which the scaffold is carried.

(6) Where counterweights are used with outriggers the counterweights shall be securely attached to the outriggers and shall be not less in weight than three times the weight which would counter-balance the weight suspended from the outrigger including the weight of the runway, joist or rail track, the suspended scaffold and persons and other loads thereon.

(7) The points of suspension of every suspended scaffold shall be an adequate horizontal distance from the face of the building or other structure.

(8) Every runway, joist and rail track supporting a suspended scaffold shall be of suitable and sound material, adequate strength for the purpose for which it is used and free from patent defect, shall be provided with adequate stops at each end and shall be properly secured to the building or other structure or, where outriggers are used, to the outriggers.

(9) The suspension ropes or chains of a suspended scaffold –

(a) shall be securely attached to the outriggers or other supports and to the platform framework or to any lifting appliance or other device attached thereto, as the case may be; and

(b) shall be kept in tension.

(10) Where winches are used with suspended scaffolds the suspension ropes shall be of such length that at the lowest positions at which the scaffold is intended to be used there are not less than two turns of rope remaining on each winch drum and the length of each rope shall be clearly marked on its winch.

(11) If a suspended scaffold is carried on fibre ropes and pulley blocks the ropes shall be spaced not more than 3.20 metres apart.

(12) Where the work to be carried out from a suspended scaffold is of such a light nature and the material required for the work is such that a cradle or similar light-weight suspended scaffold can be used with safety and where such suspended scaffold is used, the requirements of paragraph (5) that the outriggers shall be installed horizontally and that stops shall be provided shall not apply.

48B. In regulation 48A –

'slung scaffold' means a scaffold suspended by means of lifting gear, ropes or chains or rigid members and not provided with means of raising or lowering by a lifting appliance or similar device;

'suspended scaffold' means a scaffold (not being a slung scaffold) suspended by means of ropes or chains and capable of being raised or lowered but does not include a boatswain's chair or similar appliance.'.

The Health and Safety (Enforcing Authority) Regulations 1989

4. In regulation 2 of the Health and Safety (Enforcing Authority) Regulations 1989, in paragraph (1) for the definitions of 'construction work' and 'contractor', substitute –

' "construction work" and "contractor" have the meanings assigned to them by regulation 2(1) of the Construction (Design and Management) Regulations 1994 [S.I. 1994/3140];'.

The Workplace (Health, Safety and Welfare) Regulations 1992

5. The following sub-paragraph shall be substituted for sub-paragraph (b) of regulation 3(1) of the Workplace (Health, Safety and Welfare) Regulations 1992 –

'(b) a workplace where the only activity being undertaken is construction work within the meaning assigned to that phrase by regulation 2(1) of the Construction (Health, Safety and Welfare) Regulations 1996 [S.I. 1996/], except for any workplace from which the application of the said Regulations is excluded by regulation 3(2) of those Regulations;'.

The Management and Administration of Safety and Health at Mines Regulations 1993

6. Paragraph (3) of regulation 40 of the Management and Administration of Safety and Health at Mines Regulations 1993 shall be deleted and the following substituted –

'(3) Those provisions of the 1961 Act which are applied by section 127 of that Act to building operations shall apply, as respects premises forming part of a mine, to building operations undertaken above ground, and in this paragraph the term "building operation" has the same meaning as in section 176(1) of the 1961 Act.'.

The Railways (Safety Case) Regulations 1994

7. In regulation 2(1) of the Railways (Safety Case) Regulations 1994 –

(a) the definition of 'building operation' shall be deleted and the following substituted –

' "building operation" means the construction, structural alteration, repair or maintenance of a building (including repointing, redecoration and external cleaning of the structure), the demolition of a building, and the preparation for, and laying the foundation of, an intended building, but does not include any operation which is a work of engineering construction;'.

(b) the definition of 'work of engineering construction' shall be deleted and the following substituted –

' "work of engineering construction" means the construction of any railway line or siding otherwise than upon an existing railway, and the construction, structural alteration or repair (including repointing and repainting) or demolition of any tunnel, bridge or viaduct, except where carried on upon a railway;'.

The Railways (Safety Critical Work) Regulations 1994

8. In regulation 2(1) of the Railways (Safety Critical Work) Regulations 1994 –

(a) the definition of 'building operation' shall be deleted and the following substituted –

' "building operation" means the construction, structural alteration, repair or maintenance of a building (including repointing, redecoration and external cleaning of the structure), the demolition of a building, and the preparation for, and laying the foundation of, an intended building, but does not include any operation which is a work of engineering construction;';

(b) the definition of 'work of engineering construction' shall be deleted and the following substituted –

' "work of engineering construction" means the construction of any railway line or siding otherwise than upon an existing railway, and the construction, structural alteration or repair (including repointing and repainting) or demolition of any tunnel, bridge or viaduct, except where carried on upon a railway or tramway.'.

The Construction (Design and Management) Regulations 1994

9. In regulation 2(1) of the Construction (Design and Management) Regulations 1994, the definition of 'construction work' shall be amended so that, in sub-paragraph (a) of that definition, in place of the words 'regulation 7 of the Chemicals (Hazard Information and Packaging) Regulations 1993,' there shall be substituted 'regulation 5 of the Carriage of Dangerous Goods by Road and Rail (Classification, Packaging and Labelling) Regulations 1994 [S.I. 1994/669],'.

Schedule 10, Regulation 35, Revocations

Column 1	Column 2	Column 3
Description of Instrument	**Reference**	**Extent of Revocation**
The Engineering Construction (Extension of Definition) Regulations 1960.	S.I. 1960/421.	The whole instrument.

The Construction (General Provisions) Regulations 1961.	S.I. 1961/1580.	Regulations 8 to 19, 21, 23 to 41, 45 to 51, 53 and 56.
The Construction (Working Places) Regulations 1966.	S.I. 1966/94.	The whole instrument.
The Construction (Health and Welfare) Regulations 1966.	S.I. 1966/95.	The whole instrument.
The Engineering Construction (Extension of Definition) (No. 2) Regulations 1968.	S.I. 1968/1530.	The whole instrument.
The Construction (Health and Welfare) (Amendment) Regulations 1974.	S.I. 1974/209.	The whole instrument.

Explanatory note

(This note is not part of the Regulations)

1. These Regulations impose requirements with respect to the health, safety and welfare of persons at work carrying out 'construction work', defined in regulation 2(1), and of others who may be affected by that work.

2. The Regulations replace the Construction (General Provisions) Regulations 1961, the Construction (Health and Welfare) Regulations 1966 and the Construction (Working Places) Regulations 1966.

3. The Regulations give effect as respects Great Britain to the following provisions of Council Directive 92/57/EEC (OJ No. L245, 26.8.92, p. 6) on the implementation of minimum safety and health requirements at temporary or mobile construction sites:

(a) Articles 8(a), (b) and (d), 9(a), and paragraph 1(a)(i) of Article 10 (in so far as it refers to Article 8(a), (b) and (d) and Annex IV);

(b) in part A of Annex IV, points 1.1, 1.2, sections 3 to 5, 7 to 12 and 14 to 18;

(c) in section II of part B of Annex IV, sections 1, 3 to 6, points 8.1(b) and (c), 8.2, 8.3, 8.4 (in part), 9.1(b) to (d) and sections 10 to 14.

4. Specified regulations apply only in respect of construction work carried out on a 'construction site', defined in regulation 2(1), and where a workplace on a construction site is set aside for purposes other than construction work, the Regulations do not apply.

5. Subject to specific exceptions, the Regulations impose requirements on –

(a) employers, the self-employed and others who control the way in which construction work is carried out;

(b) employees in respect of their own actions; and

(c) every person at work as regards co-operation with others and the reporting of danger.

6. The Regulations impose requirements with respect to –

(a) the provision of safe places of work and safe access and egress thereto (regulation 5);
(b) the provision of suitable equipment to prevent falls (regulation 6);
(c) the working on or near fragile material (regulation 7);
(d) the prevention of injury from falling objects (regulation 8);
(e) the stability of structures (regulation 9);
(f) the carrying out and supervision of demolition and dismantling and the use of explosives (regulations 10 and 11);
(g) the safety of excavations, cofferdams and caissons (regulations 12 and 13);
(h) the prevention of drowning (regulation 14);
(i) the movement of pedestrian and vehicular traffic (regulation 15);
(j) the construction of doors, gates and hatches (regulation 16);
(k) the use of vehicles (regulation 17);
(l) the risks from fire, the provision of emergency routes and exits, the preparation and implementation of evacuation procedures and the provision of fire-fighting equipment, fire detectors and alarms (regulations 18 to 21);
(m) the provision of sanitary and washing facilities, a supply of drinking water, rest facilities and facilities to change and store clothing (regulation 22);
(n) the provision of adequate fresh air, reasonable temperature and weather protection (regulations 23 and 24);
(o) the provision of lighting (including emergency lighting) (regulation 25);
(p) the marking and good order of a construction site (regulation 26);
(q) the safety and maintenance of plant and equipment (regulation 27);
(r) training and supervision (regulation 28);
(s) the inspection of places of work and the preparation of reports (regulations 29 and 30).

7. Regulation 31 provides for the granting of exemptions from the Regulations by the Health and Safety Executive.

8. Regulation 32 extends the application of the Regulations to construction activities within territorial waters.

9. Regulation 33 provides for specified provisions of the Regulations to be enforced in specified circumstances by fire authorities.

10. Regulation 34 modifies the enactments referred to in Schedule 9, and regulation 35 revokes the enactments referred to in Schedule 10.

11. A copy of the cost benefit assessment prepared in respect of these Regulations can be obtained from the Health and Safety Executive, Construction Policy Section, Safety Policy Directorate, Rose Court, 2 Southwark Bridge, London SE1 9HS. A copy has been placed in the Library of each House of Parliament.

Proposed changes to the CDM Regulations

In 2005 the Health and Safety Commission issued a consultative document for the construction industry which proposed a significant realignment of construction health and safety legislation. It is widely anticipated that in April 2007 a revised regulation will appear. It is expected to retain the name of the Construction (Design and Management) Regulations.

Taking its thrust from a 2002 discussion document entitled *Revitalising Health and Safety in Construction*, other industry feedback and a run of very poor health and safety statistics, the Commission set about proposals for a regulatory makeover.

The review has built on the underlying principles of the existing CDM Regulations and aims to 'simplify and clarify' what the five existing duty holders need to do to carry out their respective roles more effectively.

The review also unites the existing CDM Regulations and the Construction (Health, Safety and Welfare) Regulations 1996. This will bring together the two sets of regulations that emerged as a result of the original EU directive.

The responses received from *Revitalising Health and Safety in Construction* endorsed the basic principles of the CDM Regulations, but suggested that a number of key issues needed addressing:

- improving competence across all duty holders;
- appreciation by clients of the influence they hold to set standards;
- assessing the effectiveness of the planning supervisor;
- clarifying the legislation so that everyone knows exactly what is expected of them;
- improving consultation with the workforce.

The Commission took this information on board, drew up a new set of proposals and issued the consultative document actively promoting comments from across the construction industry.

The following summary was issued by the Health and Safety Executive during early 2006:

Key CDM changes – proposed following responses to the Consultative Document

Clients

1. The duties on designers and contractors to check that clients are aware of their duties should be extended to non-notifiable projects.
2. We want to be clear that clients need to take **reasonable steps** to ensure that there are **reasonable** management arrangements for the project, designed to ensure **reasonable** standards of health and safety. We do not expect them, particularly those who have virtually no knowledge of the construction industry, to ensure the arrangements achieve zero risk!
3. The duty on clients to provide information should:

    *   extend to non-notifiable projects
    *   make it clear that information which is reasonably obtainable includes, when necessary, that which can be obtained through surveys and by similar investigation, when necessary.

4. The duty to update a health and safety file should only extend to information discovered during, or arising from, the project. Clients/ project teams don't have to seek out unrelated information to create a file for the whole structure or site.

Co-ordinator

5. The distinction between duties and functions puzzled a lot of people, as did the possibility of having more than one co-ordinator. We propose to drop the reference to functions and to have only one co-ordinator, while still giving clients/co-ordinators the option to delegate much of the actual work to a third party. The co-ordinator's duties would be to:

    *   advise and assist the client;
    *   notify HSE or the Office of the Rail Regulator (for work on the railways) along the lines currently set out in CDM94 reg 7(3);
    *   ensure, so far as reasonably practicable, that suitable arrangements are made and implemented for –
        *   identifying/obtaining the information specified in regulation 10 (in co-operation with whoever has it or can commission surveys, etc. to obtain it) and appropriately and promptly communicating it (directly or indirectly);
        *   co-ordination and co-operation between design businesses of the process of risk avoidance/management required by regulation 14;
        *   planning and other preparation for the construction phase – but only until the PC takes up the reins;

- co-ordination and co-operation during the construction phase between designers and the principal contractor in relation to any design or change to a design requiring a review of the construction phase plan;
- the preparation, where none exists, and otherwise review and updating of the health and safety file and its delivery to the client at the end of the construction phase.

(This revised regulation will need to be compared/contrasted with reg 7 to see if there is scope to rationalise or simplify them.)

Design

6. We want to be clear that designers are only responsible for their own designs, not necessarily for the design of the whole structure.
7. We want to allow some flexibility in the timing of the appointment of the co-ordinator. Some outline design work can be done before the appointment is made, but we still want to ensure that the co-ordinator is in place before any design decisions become 'set in concrete'. We have not identified a clear way of writing this into the regulations and may have to clarify it in the ACoP.
8. We need to make it clear that the scope of draft regulation 14(3)(c) only extends to matters of HSE interest – e.g. not fire safety. We want the completed structure to comply with health and safety law in general and the Workplace Regs in particular.
9. We propose to add an additional requirement to make it explicit that designers must provide the information needed for the health and safety file.

Principal contractor and contractors

10. We propose to add a requirement to implement the health and safety plan, so far as reasonably practicable.
11. In draft regulation 19(5)(d) we want to make it clear that we expect the PC and contractor to work together to address any problems with the plan.

Notification

12. Notification should be a duty of the co-ordinator, rather than the client.
13. To avoid lots of abortive notifications we will revert to the current CDM timing for notification.
14. We propose to add details of any lead designer, the minimum mobilisation period and a declaration from the client that he is aware of his

duties under the regulations to the notification requirements in Schedule 1.

Transitional provisions

15. Rather than assuming that the planning supervisor and principal contractor automatically take on the new/modified roles we propose to make it clear that they must agree to do so.
16. An agent appointed under reg 4 of CDM 1994, prior to the date of enactment of these regs may be treated (subject to their agreement) as the client under the new regulations until the end of the project or for 5 years after they come into force, whichever is sooner.

Schedules 2 and 3

17. We propose to make these separate parts of the regulations as some felt that including them in schedules made it appear that they were less important. We propose to add a requirement to provide a means of heating for welfare facilities.
18. There are also quite a lot of other minor drafting changes that need to be made to make the drafting clearer.

Glossary of terms

ACoP   See *Approved Code of Practice*.

Agent   A person whose trade, business or other undertaking (whether for profit or not) is to act as an agent for a client. Employees of the client who discharge functions on behalf of the client are not agents of the client for the purposes of these definitions.

Approved Code of Practice (ACoP)   Health and safety codes of practice are approved by the Health and Safety Commission and have special legal status. The code associated with the CDM Regulations is contained within *Managing Health and Safety in Construction: Construction (Design and Management) Regulations 1994* (ref. HSG224, by HSE Books, referred to as the ACoP). See also Section 1, 'Overview of the CDM Regulations'.

Cleaning work   The cleaning of any window or any transparent or translucent wall, ceiling or roof in or on a structure where it involves a risk of a person falling more than 2 metres (see regulation 2 of CDM).

Client   In the ACoP, client includes an agent, if one has been appointed.

Client's agent   A 'client's agent' for the purposes of CDM is any agent or other client appointed to act as the only client. The appointment may be 'declared' to the HSE, or not (based on regulation 4).

Competence   Having, or having ready access to, the skills, knowledge, experience, systems and support necessary to carry out work relating to the construction work in hand, in a manner that takes due account of health and safety issues.

Construction phase   'The period pf time commencing when construction work in any project starts and ending when construction work in that project is completed' (regulation 2(1)).

Construction phase health and safety plan   Arrangements for the management of the construction work to ensure the health and safety of all those involved or affected by the work.

Construction risk assessment   The process of risk assessment (see *Risk assessment*) applied by contractors to the construction process.

Construction work   'The carrying out of any building, civil engineering or engineering construction work and includes any of the following:

- the construction, alteration, conversion, fitting out, commissioning, renovation, repair, upkeep, redecoration or other maintenance (including cleaning which involves the use of water or an abrasive at high pressure or the use of substances classified as corrosive or toxic for the purposes of Regulation 7 of the Chemicals (Hazard Information and Packaging) Regulations 1993), de-commissioning, demolition or dismantling of a structure,
- the preparation for an intended structure, including site clearance, exploration, investigation (but not site survey) and excavation, and laying or installing the foundations of the structure,
- the assembly of prefabricated elements to form a structure or the disassembly of prefabricated elements which, immediately before such disassembly, formed a structure, and
- the removal of a structure or part of a structure or of any product or waste resulting from demolition or dismantling of a structure or from disassembly of prefabricated elements which, immediately before such disassembly, formed a structure, and
- the installation, commissioning, maintenance, repair or removal of mechanical, electrical, gas, compressed air, hydraulic, telecommunications, computer or similar services, which are normally fixed within or to a structure,

but does not include the exploration for or extraction of mineral resources or activities preparatory thereto carried out at a place where such exploration or extraction is carried out' (regulation 2(1)).

Contractor   An organisation or individual who carries on a trade, business or other undertaking in connection with which they undertake, carry out or manage construction work – includes subcontractors. Regulation 2(1) of CDM contains the full definition.

Control   Measure taken to reduce (or mitigate) risk.

Demolition/dismantling   The deliberate pulling down, destruction or taking apart of a structure, or a substantial part of a structure. It includes dismantling for re-erection or reuse. Demolition does not include operations such as making openings for doors and windows, or services for removing non-structural elements such as cladding, roof tiles or scaffolding. These operations may, however, form part of demolition or dismantling work when carried out alongside other activities.

Design   'Design in relation to any structure includes drawings, design details, specifications and bill of quantities (including specifications of articles or substances) in relation to the structure' (regulation 2(1)).

Designer   'Any person who carries out a trade, business or other undertaking in connection with which he prepares a design relating to a structure or part of a structure' (regulation 2(1)). This will include not only design professionals but also others who make decisions about materials or how construction work will be done.

Design risk assessment   The process of risk assessment (see *Risk assessment*) applied by designers to their design.

Developer   For the purpose of the CDM Regulations, a developer has a special meaning (see regulation 5) 'when commercial developers sell domestic premises before the project is complete and arrange for construction work to be carried out', whereby the regulations apply as if the developer were client and the work is not treated as being for a domestic client. In other words, a commercial housing developer is the client for the housing development even though he may have sold some or all of the properties before completion of the works.

Domestic client   A client for whom a project is carried out which is not related to the client trade or business (whether for profit or not).

Duty holder   Someone who has duties under CDM.

Fixed plant   Fixed plant means (for the purpose of CDM) plant and machinery that is used for a process; it is not part of the services associated with the structure of the building services.

Fragile material   A surface assembly liable to fail from the weight of anyone crossing, working or falling on it (including the weight of anything that they may be carrying). Any surface or assembly may be fragile, particularly if incorrectly fixed, supported or specified. All tend to deteriorate with age, exposure to UV light and weathering. Typical fragile materials are roof lights, fibre cement sheets, corroded metal sheets, glass (including wired glass) and wood wool slabs. They present a risk to people installing the material, doing subsequent maintenance and crossing them to gain access to other parts of the structure, or plant situated on the roof.

Hazard   Something with the potential to cause harm (this can include articles, substances, plant or machines, methods of work, the working environment and other aspects of work organisation).

Health and safety file   Information which people, including clients, designers, planning supervisors, contractors and others involved in carrying out construction or cleaning work on the structure in the future are likely to need, but could not be expected to know.

Health and safety plan   This is a document that contains information to assist with the management of heath and safety and the project proceeds. It has two main stages, pre-tender and construction. The pre-tender health and safety plan (so named because it is normally prepared before the tendering process for the construction contract) brings together the health and safety information obtained for the client and designers. The construction health and safety plan details how the construction work will be managed on site to ensure health and safety.

HSC   Health and Safety Commission.

HSE   Health and Safety Executive.

Maintenance The repair, renovation, upkeep, redecoration and high-pressure cleaning with water or abrasives, or cleaning with corrosive or toxic substances of structures. The maintenance of services that are normally fixed to or within a structure is covered by CDM, but the maintenance of other fixed plant is not covered. The definitions of construction work and structure in regulation 2 provide more detail.

Method statement A written document laying out work procedures and sequence of operation. It take account of the risk assessment carried out for the task or operation and the control measures identified.

Notifiable Construction work is notifiable to the HSE if CDM applies and if the construction phase (including commissioning) is expected to last more that 30 working days or will involve more that 500 person days of work.

Planning supervisor The regulations define the planning supervisor as a person who carries out the function as defined. The planning supervisor is the person who coordinates and manages the health and safety aspects of design. The planning supervisor also has to ensure that the pre-tender stage of the health and safety plan and the health and safety file are prepared.

Practicable This indicates a high standard but not an absolute one.

Pre-tender health and safety plan A plan containing the information required by regulation 15(3) of CDM, including information obtained from the client and designer, during the design and the early planning stages, for use by contractors, e.g. when preparing an offer to the client.

Principal contractor A role prescribed by the regulations. The principal contractor is a contractor who is appointed by the client. The principal contractor has the overall responsibility for the management and coordination of site operations with respect to health and safety.

Project This means 'a project that includes, or is intended to include, construction work' (regulation 2(1)).

Reasonably practicable This indicates a somewhat lesser standard than 'practicable', with a more balanced approach permissible.

Residual hazards/risks The hazards/risks that remain after the design process.

Risk The likelihood of potential harm from a hazard being realised. The extent of the risk depends on: (1) the likelihood of that harm occurring; (2) the potential severity of that harm, i.e. of any resultant injury or adverse health effect; and (3) the population which might be affected by the hazard, i.e. the number of people who might be exposed. TheACoP to regulation 3 of the Management Regulations provides further details.

Risk assessment The process of identifying hazards, assessing the degree of risk associated with them and identifying suitable control measures.

Shall In legal terms this indicates an absolute obligation to undertake the act.

So far as reasonably practicable To carry out a duty 'so far as reasonably practicable' means that the degree of risk in a particular activity can be balanced against the time, trouble, cost and physical difficulty of taking measures to avoid the risk. If these are so disproportionate to the risk that it would be quite unreasonable for the people concerned to have to incur them to prevent it, they are not obliged to do so. The greater the risk, the more likely it is that it is reasonable to go to very substantial expense, trouble and intervention to reduce it. However, if the consequences and the extent of a risk are small, insistence on great expense will not be considered reasonable.

Structure   '(a) any building, steel or reinforced concrete structure (not being a building,) railway line or siding, tramway line, clock, harbour, inland navigation, tunnel, shaft, bridge, viaduct, waterworks, reservoir, pipe or pipe-line (whatever, in either case, it contains or is intended to contain), cable, aqueduct, sewer, sewage works, gasholder, road, airfield, sea defence works, river works, drainage works, earthworks, lagoon, dam, wall, caisson, mast, tower, pylon, underground tank, earth retaining structure, or structure designed to preserve or alter any natural feature, and any other structure similar to the foregoing, or

(b) any formwork, falsework, scaffold or other structure designed or used to provide support or means of access during construction work, or

(c) any fixed plant in respect of work which is installation, commissioning, de-commissioning or dismantling and where any such work involves a risk of a person falling more than 2 metres.'

Appendix 5

References and further reading

References

Accelerating Change, Strategic Forum for Construction (chaired by Sir John Egan), 2002, Rethinking Construction c/o Construction Industry Council, London, http://strategicforum.org.uk/pdf/report_sept02.pdf, ISBN 18986 71281

Backs for the Future: Safe Manual Handling in Construction, HSG149, HSE Books, 2000, ISBN 07176 11221

CDM Regulations Procedures Manual, second edition, Stuart Summerhayes, 2002, ISBN 1405 107405

CDM Regulations: Work Sector Guidance for Designers, CIRIA Report 166 and C604, Construction Industry Research and Information Association (CIRIA), 1997, ISBN 086017 4646

Fire Safety in Construction Work, HSG168, HSE Books, 2001, ISBN 07176 13321

Health and Safety in Excavations: Be Safe and Shore, HSG185, HSE Books, 1999, ISBN 07176 15634

Health and Safety in Roof Work, HSG33, HSE Books, 1999, ISBN 07176 14255

Management of Health and Safety at Work: Management of Health and Safety at Work Regulations 1999. Approved Code of Practice and Guidance, second edition, L21, HSE Books, 2000, ISBN 07176 24889

Managing Health and Safety in Construction, HSG224, HSE Books, 1994, ISBN 07176 21391

Protecting the Public: Your Next Move, HSG151, HSE Books, 2001, ISBN 07176 11485

Safe Access for Maintenance and Repair, CIRIA C611, CIRIA, 2003, ISBN 086017 6118

Successful Health and Safety Management, second edition, HSG65, HSE Books, 2000, ISBN 07176 12767

The Safe Use of Vehicles on Construction Sites, HSG144, HSE Books, 2003, ISBN 07176 1610x

Further reading

A Guide to the Construction (Health, Safety and Welfare) Regulations 1996, revised edition, INDG220, HSE Books, 2001, ISBN 0 7176 1161 2

CDM Regulations: Practical Guidance for Clients and Clients' Agents, Report 172, CIRIA, 1998, ISBN 0 86017 4867

CDM Regulations: Practical Guidance for Planning Supervisors, Report 173, CIRIA, 1998, ISBN 0 86017 487 5

Health and Safety in Construction, revised edition, HSG150, HSE Books, 2001, ISBN 0 7176 2106 5

Index

Learning Resources
Centre